HOW INNOVATION WORKS

By the same author

The Red Queen:
Sex and the Evolution of Human Nature

The Origins of Virtue

Genome:
The Autobiography of a Species in 23 Chapters

Nature via Nurture:
Genes, Experience and What Makes Us Human

Francis Crick:
Discoverer of the Genetic Code

The Rational Optimist: How Prosperity Evolves

The Evolution of Everything

HOW INNOVATION WORKS

MATT RIDLEY

Withdrawn from Stock

4th ESTATE · London

4th Estate
An imprint of HarperCollins*Publishers*
1 London Bridge Street
London SE1 9GF
www.4thEstate.co.uk

First published in Great Britain by 4th Estate in 2020

1

Copyright © Matt Ridley 2020

The right of Matt Ridley to be identified as
the author of this work has been asserted by him in accordance
with the Copyright, Design and Patents Act 1988

A catalogue record for this book is available from the British Library

ISBN HB 978-0-00-833481-9
ISBN TPB 978-0-00-833907-4

All rights reserved. No part of this publication may be reproduced,
transmitted or stored in a retrieval system, in any form or by any means,
electronic, mechanical, photocopying, recording or otherwise,
without the prior permission of the publishers.

This book is sold subject to the condition that it shall not, by way
of trade or otherwise, be lent, re-sold, hired out or otherwise circulated
without the publisher's prior consent in any form of binding or cover other
than that in which it is published and without a similar condition including
this condition being imposed on the subsequent purchaser.

Typeset in Sabon and Avenir

Printed and bound in in Great Britain by
CPI Group (UK) Ltd, Croydon, CR0 4YY

MIX
Paper from
responsible sources
FSC™ C007454

This book is produced from independently certified FSC paper
to ensure responsible forest management

Find out more about HarperCollins and the environment at
www.harpercollins.co.uk/green

For Felicity Bryan

CONTENTS

INTRODUCTION

The Infinite Improbability Drive

Innovation offers the carrot of spectacular reward
or the stick of destitution.

JOSEPH SCHUMPETER

I am walking along a path on the Inner Farne, an island off the coast of north-east England. By the side of the path, amid the sea-campion flowers, sits a female eider duck, dark brown and broody, silently incubating her clutch of eggs. I stoop to take a picture of her with my iPhone from a few feet away. She is used to this: hundreds of visitors come here every day in summer and many will take her picture. For some reason, an idea pops into my head as I click: a riff on the second law of thermodynamics based on a remark by my friend John Constable. The idea is this: the electricity in the iPhone's battery and the warmth in the eider duck's body are doing roughly the same thing: making improbable order (photographs, ducklings) by expending or converting energy. And then I think that the idea I've just had itself, like the eider duck and the iPhone, is also an improbable arrangement of synaptic activity in my brain, also fuelled by energy from the food I have recently eaten, of course, but made possible by the underlying order of the brain, itself the evolved product of millennia of natural selection acting on individuals, each of whose own improbabilities were sustained by energy conversion. Improbable arrangements of the world, crystallized consequences of energy generation, are what both life and technology are all about.

In Douglas Adams's *The Hitchhiker's Guide to the Galaxy*, Zaphod Beeblebrox's starship *Heart of Gold* – a metaphor for wealth – is powered by a fictional 'infinite improbability drive'. Yet a near-infinite improbability drive does indeed exist, but here on Planet Earth, in the shape of the process of innovation. Innovations come in many forms, but one thing they all have in common, and which they share with biological innovations created by evolution, is that they are enhanced forms of improbability. That is to say, innovations, be they iPhones, ideas or eider ducklings, are all unlikely, improbable combinations of atoms and digital bits of information. It is astronomically improbable that the atoms in an iPhone would be neatly arranged by chance into millions of transistors and liquid crystals, or the atoms in an eider duckling would be arranged to form blood vessels and downy feathers, or the firings of neurons in my brain would be arranged in such a pattern that they can and sometimes do represent the concept of 'the second law of thermodynamics'. Innovation, like evolution, is a process of constantly discovering ways of rearranging the world into forms that are unlikely to arise by chance – and that happen to be useful. The resulting entities are the opposite of entropy: they are more ordered, less random, than their ingredients were before. And innovation is potentially infinite because even if it runs out of new things to do, it can always find ways to do the same things more quickly or for less energy.

In this universe it is compulsory, under the second law of thermodynamics, that entropy cannot be reversed, locally, unless there is a source of energy – which is necessarily supplied by making something else even less ordered somewhere else, so the entropy of the whole system increases. The power of the improbability drive is therefore limited only by the supply of energy. So long as human beings apply energy to the world in careful ways, they can create ever more ingenious and improbable structures. The medieval castle at Dunstanburgh I can see from the island is an improbable structure, and its partial ruin after 700 years is more probable, more entropic. The castle in its prime was the direct consequence of the expenditure of lots of energy, in this case mainly in the muscles of masons who were fed with bread and cheese that was made from wheat and grass that was grown in sunlight and eaten by cows. John Constable, a former Cambridge and Kyoto academic, points out that the things we rely on to make our lives prosperous are

all of them, without exception, physical states far from thermodynamic equilibrium, and the world was brought, sometimes over long periods of time, into these convenient configurations by energy conversion, the use of which reduced entropy in one corner of the universe, ours, and increased it by an even larger margin somewhere else. The more ordered and improbable our world becomes, the richer we become, and, as a consequence, the more disordered the universe becomes overall.

Innovation, then, means finding new ways to apply energy to create improbable things, and see them catch on. It means much more than invention, because the word implies developing an invention to the point where it catches on because it is sufficiently practical, affordable, reliable and ubiquitous to be worth using. The Nobel Prize-winning economist Edmund Phelps defines an innovation as 'a new method or new product that becomes a new practice somewhere in the world'. In the pages that follow I will trace the path of ideas from the invention to the innovation, through the long struggle to get an idea to catch on, usually by combining it with other ideas.

And here is my starting point: innovation is the most important fact about the modern world, but one of the least well understood. It is the reason most people today live lives of prosperity and wisdom compared with their ancestors, the overwhelming cause of the great enrichment of the past few centuries, the simple explanation of why the incidence of extreme poverty is in global freefall for the first time in history: from 50 per cent of the world population to 9 per cent in my lifetime.

What made most of us, not just in the West but in China and Brazil too, unprecedentedly rich, so the economic historian Deirdre McCloskey says, was 'innovationism': the habit of applying new ideas to raising living standards. No other explanation of the great enrichment of recent centuries makes any sense. Trade had been expanding for centuries, and colonial exploitation with it, and these alone were unable to give anything like the order of magnitude of improvement in incomes that happened. There was no sufficient accumulation of capital to make such a difference, no 'piling of brick on brick, or bachelor's degree on bachelor's degree' in McCloskey's words. There was no sufficiently great expansion in the availability of labour. Nor was the scientific revolution of Galileo and Newton responsible, for most of the innovations that changed people's lives at least at first owed little to new scientific knowledge and few of the innovators who drove the changes were trained scientists. Indeed many, such as Thomas Newcomen, the inventor of the steam engine, or Richard Arkwright of the textile revolution, or George Stephenson of the railways, were poorly educated men of humble origins. Much innovation preceded the science that underpinned it. The Industrial Revolution therefore was in effect, as Phelps has argued, the emergence of a new kind of economic system that generated endogenous innovation as a product in itself. I will argue that some machines themselves made this possible. A steam engine proved to be 'autocatalytic': it drained the mines, which cut the cost of coal, which made the next machine cheaper and easier to make. But I am getting ahead of myself.

The word 'innovation' is invoked with alarming frequency by companies trying to sound up to date but with little or no systematic idea about how it occurs. The surprising truth is that nobody really knows why innovation happens and how it happens, let alone when and where it will happen next. One economic historian, Angus Maddison, wrote that 'technical progress is the most essential characteristic of modern growth and one that is most difficult to quantify or explain'; another, Joel Mokyr, said that scholars 'know remarkably little about the kind of institutions that foster and stimulate technological progress'.

Take sliced bread, for example. Best thing since, and all that. Looking back it is obvious that somebody would invent a way of automatically pre-slicing bread to make uniform sandwiches. It is fairly obvious that this would probably happen in the first half of the twentieth century when electrical machines were all the rage for the first time. But why 1928? And why in the small town of Chillicothe, in the middle of Missouri? Lots of people tried to make bread-slicing machines, but they either worked poorly or they led to stale bread because it was not well packaged. The person who made it work was Otto Frederick Rohwedder, who was born in Iowa, was educated as an optician in Chicago and set up shop as a jeweller in St Joseph, Missouri, before moving back to Iowa determined – for some reason – to invent a bread slicer. He lost his first prototype in a fire in 1917 and had to start all over again. Crucially he realized that he must invent automatic packaging of the bread at the same time lest the slices go stale. Most bakeries were not interested, but the Chillocothe bakery, owned by one Frank Bench, was and the rest is history. What was special about Missouri? Beyond a general mid-twentieth-century American affection for innovation and the means to make it happen, the best guess is that it was a slice of random luck. Serendipity plays a big part in innovation, which is why liberal economies, with their free-roving experimental opportunities, do so well. They give luck a chance.

Innovation happens when people are free to think, experiment and speculate. It happens when people can trade with each other. It happens where people are relatively prosperous, not desperate. It is somewhat contagious. It needs investment. It generally happens in cities. And so on. But do we really understand it? What is the best way to encourage innovation? To set targets, direct research, subsidize science, write rules and standards; or to back off from all this, deregulate, set people free; or to create property rights in ideas, offer patents and hand out prizes, issue medals; to fear the future; or to be full of hope? You will find champions of all these policies and more, fervently arguing their cases. But the striking thing about innovation is how mysterious it still is. No economist or social scientist can fully explain why innovation happens, let alone why it happens when and where it does.

In this book I shall try to tackle this great puzzle. I will do so not by abstract theorizing or argument alone, though there will be some of both, but mainly by telling stories. Let the innovators who turned their (or other people's) inventions into useful innovations teach us, by the examples of their successes and failures, how it happened. I tell the stories of steam engines and search engines, of vaccines and vaping, of shipping containers and silicon chips, of wheeled suitcases and gene editing, of numbers and water closets. Let's hear from Thomas Edison and Guglielmo Marconi, from Thomas Newcomen and Gordon Moore, from Lady Mary Wortley Montagu and Pearl Kendrick, from Al Khwarizmi and Grace Hopper, from James Dyson and Jeff Bezos.

I cannot hope to document every important innovation. I have omitted some very important and well-known ones for no particular reason: the automation of the textile industry, for example, or the history of the limited company. I have left out most innovation in art, music and literature. My main examples are drawn from the worlds of energy, public health, transport, food, low technology, and computers and communications.

Not all the people whose stories I tell are heroes; some are frauds, fakers or failures. Few worked alone, for innovation is a team sport, a collective enterprise, far more than is generally recognized. Credit and authorship are confused and mysterious if not downright unfair. Yet unlike most team sports innovation is not usually a choreographed, planned or managed thing. It cannot be easily predicted, as many a red-faced forecaster has discovered. It runs mostly on trial and error, the human version of natural selection. And it usually stumbles on great breakthroughs when looking for something else: it is heavily serendipitous.

I will plunge back in time to the very start of human culture to try to understand what triggered innovation in the first place and why it happens to people but not to robins or rocks. Chimpanzees and crows do innovate, by developing and spreading new cultural habits, but very occasionally and rather slowly; most other animals not at all.

In the ten years since I published *The Rational Optimist*, arguing unfashionably that the world has been, is, and will go on getting better, not worse, human living standards have grown rapidly higher for nearly everybody. I finished that book as the world was plumbing the depths of a terrible recession, but the years since have been ones of faster economic growth for much of the poor of the world than ever before. The income of the average Ethiopian has doubled in a decade; the number of people living in extreme poverty has dipped below 10 per cent for the first time in history; malaria mortality has plummeted; war has ceased altogether in the western hemisphere and become much rarer in the Old World, too; frugal LED lights have replaced both incandescent and fluorescent bulbs; telephone conversations have essentially become free on Wi-Fi. Some things have got worse, of course, but most trends are positive. All this is due to innovation.

The chief way in which innovation changes our lives is by enabling people to work for each other. As I have argued before, the main theme of human history is that we become steadily more specialized in what we produce, and steadily more diversified in what we consume: we move away from precarious self-sufficiency to safer mutual interdependence. By concentrating on serving other people's needs for forty hours a week – which we call a job – you can spend the other seventy-two hours (not counting fifty-six hours in bed) drawing upon the services provided to you by other people. Innovation has made it possible to work for a fraction of a second so as to be able to afford to turn on an electric lamp for an hour, providing the quantity of light that would have required a whole day's work if you had to make it yourself by collecting and refining sesame oil or lamb fat to burn in a simple lamp, as much of humanity did in the not so distant past.

Most innovation is a gradual process. The modern obsession with disruptive innovation, a phrase coined by the Harvard professor Clayton Christensen in 1995, is misleading. Even when a new technology does upend an old one, as digital media has done to newspapers, the effect begins very slowly, gathers pace gradually and works by increments, not leaps and bounds. Innovation often disappoints in its early years, only to exceed expectations once it gets going, a phenomenon I call the Amara hype cycle, after Roy Amara, who first said that we underestimate the impact of innovation in the long run but overestimate it in the short run.

Perhaps the most puzzling aspect of innovation is how unpopular it is, for all the lip service we pay to it. Despite the abundant evidence that it has transformed almost everybody's lives for the better in innumerable ways, the kneejerk reaction of most people to something new is often worry, sometimes even disgust. Unless it is of obvious use to ourselves, we tend to imagine the bad consequences that might occur far more than the good ones. And we throw obstacles in the way of innovators, on behalf of those with a vested interest in the status quo: investors, managers and employees alike. History shows that innovation is a delicate and vulnerable flower, easily crushed underfoot, but quick to regrow if conditions allow.

This strange phenomenon of innovation, and the resistance to it, was eloquently celebrated more than three centuries ago, before the start of the great enrichment, by an innovator – though he would not have used that word. William Petty went from being a teenage cabin boy on a ship who was marooned on a foreign shore with a broken leg, to getting a Jesuit education and becoming secretary to the philosopher Thomas Hobbes. Then, following a spell in Holland, he began a career as a physician and scientist, before emerging as a merchant, an Irish land speculator, a Member of Parliament, then a wealthy and politically influential pioneer of the study of economics. He was a better innovator than inventor. Early in his career, while a professor of anatomy in Oxford in 1647, Petty invented and patented a double-writing instrument – by which he could produce two copies of the first chapter of Hebrews in one go, in fifteen minutes – as well as a scheme for making a bridge with no supports on the river bed, and an engine for planting corn. None of them seemed to catch on. With feeling, Petty later wrote this lament about the lot of the inventor, in 1662:

Few new inventions were ever rewarded by a monopoly; for although the inventor, oftentimes drunk with the opinion of his own merit, thinks all the world will encroach and invade upon him, yet I have observed that the generality of men will scarce be hired to make use of the new substances which themselves have not thoroughly tried, and which length of time hath not vindicated from latent inconvenience, so as when a new invention is first propounded in the beginning every man objects and the poor inventor runs the gauntloop of all petulant wits, every man finding his several flaw, no man approving it unless mended according to his own device. Now, not one of a hundred outlives this torture, and those that do are at length so changed by the various contrivances of others, that not any one man can pretend to the invention of the whole, nor well agree about their respective share in the parts. And moreover this commonly is so long adoing, that the poor inventor is either dead, or disabled by the debts contracted to pursue his design; and withal railed upon as a projector or worse, by those who joined their money in partnership with his wit; so as the said inventor and his pretences are wholly lost as vanished.

1

Energy

Whenever you see a successful business, someone once made a courageous decision.

PETER DRUCKER

Of heat, work and light

Possibly the most important event in the history of human-kind, I would argue, happened somewhere in north-west Europe, some time around 1700, and was achieved by somebody or somebodies (probably French or English) – but we may never know who. Why so vague? At the time nobody would have noticed or realized its significance; and innovation was anyway a little-valued thing. There is confusion too about whose contribution among several candidates mattered most. And it was a gradual, stumbling change, with no eureka moment. These features are typical of innovation.

The event I am talking about is the first controlled conversion of heat to work, the key breakthrough that made the Industrial Revolution possible if not inevitable and hence

led to the prosperity of the modern world and the stupen-
dous flowering of technology today. (Here I use the word
'work' in its more colloquial sense, as controlled and ener-
getic movement, rather than in the broader way physicists
define it.) I am writing this on a laptop powered by electricity aboard a train also powered by electricity, and with
the help of electric light. Most of that electricity is coming
down wires from a power station in which enormous turbines are being spun at high speed by steam generated by
the burning of gas or boiled by the heat of nuclear fission.
The purpose of a power station is to turn the heat of combustion into the pressure of water expanding into steam
and thence into the movement of the blades of the turbine,
which moves inside an electromagnet to create the movement of electrons in wires. Something similar happens
inside the engine of a car or a plane: combustion causes
pressure, which causes movement. Virtually all the gigantic
amounts of energy that go into making my life and yours
happen come from the conversion of heat to work.

Before 1700 there were two main kinds of energy used
by human beings: heat and work. (Light came mainly from
heat.) People burned wood or coal to keep warm and cook
food; and they used their muscles, or those of horses and
oxen, or rarely a water wheel or a windmill, to move things,
to do work. These two kinds of energy were separate: wood
and coal did no mechanical work; wind, water and oxen
did no warming.

A few years later, albeit initially on a small scale, steam
was turning heat into work, and the world would never be
the same again. The first practical device for doing this was
the Newcomen engine, and Thomas Newcomen therefore
is my first and most promising candidate for the innovator

of the heat-to-work transition. Notice I do not call him an inventor; the difference is crucial.

We possess no portrait of Newcomen, and he is buried in an unmarked grave somewhere in Islington, north London, where he died in 1729. Not far away, though again we do not know where, lies the unmarked grave of one of his rivals and a possible source of his inspiration, Denis Papin, who simply faded from view around 1712 as a pauper in London. Only slightly more favourably treated by his own world was Thomas Savery, who died in 1715 in nearby Westminster. These three men, neighbours for a few years and near contemporaries (Papin was born in 1647, Savery probably around 1650 and Newcomen in 1663), all played crucial roles in the heat-to-work transition. But they may never have met.

They were not the first to notice that steam has the power to move things, of course. Toys built to exploit this principle were used in ancient Greece and Rome, and from time to time throughout the centuries clever engineers would build devices to use steam to push water about for fountains in gardens or some such trick. But it was Papin who first began to dream of harnessing this power for practical purposes rather than entertainment, Savery who turned a similar dream into a machine, albeit one that proved impractical, and Newcomen who made a practical machine that actually made a difference.

Or so goes the conventional narrative. Dig deeper and it gets more confusing. Was the French Papin robbed by one or both the Britons? Did Savery or Newcomen pinch his insights from the other? Was Papin perhaps inspired by Savery as much as the other way round? And was Newcomen even aware of the work of the other two?

Although he died in the most obscurity, Denis Papin was the star in terms of intellect and fame in his lifetime. He worked with many of the great scientists of the age. Born in Blois on the Loire, he studied medicine at university. He was recruited by the great Dutch natural philosopher and president of the Academy of Sciences in Paris, Christiaan Huygens, as one of his assistants in 1672, along with another clever young man destined for even greater renown, Gottfried Leibniz. Three years later, Papin found himself exiled in London to escape anti-Protestant persecution in Louis XIV's France.

There, presumably with an introduction from Huygens, he became Robert Boyle's assistant, working on an air pump. Robert Hooke then hired him briefly before Papin left for Venice, where he spent three years as a curator of a scientific society, before returning to London in 1684 to do the same job for the Royal Society. Somewhere along the line he invented the pressure cooker for softening bones. By 1688 he had become a professor of mathematics at the University of Marburg, before moving to Cassel in 1695. There is a sense either of restlessness or that nobody could stand his company for very long.

Huygens had employed Papin to explore the idea of a machine driven by a vacuum created by the explosion of gunpowder in a cylinder (an idea that is distantly ancestral to the internal-combustion engine), but he soon realized that the condensing of steam might work better. Some time between 1690 and 1695 he even built a simple piston and cylinder in which steam could condense on cooling, causing the piston to plunge, thereby lifting a weight by a pulley. He had discovered the principle of the atmospheric engine in which it is the weight of the atmosphere that does the work once a vacuum has been created under the piston. It is a machine that sucks rather than blows.

In the summer of 1698 Leibniz exchanged letters with Papin about the latter's designs for engines that could raise water by the use of fire. Pumping water out of mines was the chief problem to be solved, for it was the one place where horses were difficult to use and where fuel was abundant. Wet mines were safer than dry ones, because the fire risk was lower, but flooding kept foiling the miners.

Yet Papin was already dreaming of powering boats by steam: 'I believe that this invention can be used for many other things besides raising water,' he wrote to Leibniz. 'In regard to travel by water I would flatter myself to reach this goal quickly enough if I could find more support.' The idea was that steam from a boiler would push a piston ejecting water through a pipe on to a paddle wheel. The piston then returned through a combination of new water being readmitted to the piston chamber and the condensation of the steam. In 1707 Papin actually built a boat with a paddle wheel, though he does not seem to have got it working by steam, but by manpower instead, to demonstrate the superiority of paddle wheels over oars. He trundled down the River Weser in it on the way to England. The professional boatmen took umbrage at this competition and destroyed the craft: Luddites before Ludd.

The historian L. T. C. Rolt concludes that Papin could have done more than he did: 'Tantalisingly, having reached the very brink of practical success, the brilliant Papin turned aside.' He returned to steam when Leibniz told him about Thomas Savery's patent on the use of fire for raising water, a patent granted in 1698 on the very day that Papin boasted to Leibniz that he knew how to make such a machine. Papin then built a different steam engine, which, from the diagram he drew, is clearly a modified version of a Savery engine. Yet it is surely possible that Savery had heard of Papin's designs from the various letters Papin sent to former colleagues at the Royal Society, though his machine is quite distinct from Papin's. Who was copying whom?

The coincidence of timing is strange, but quite characteristic of inventors. Again and again, simultaneous invention marks the progress of technology as if there is something ripe about the moment. It does not necessarily imply plagiarism. In this case the combination of better metalworking, more interest in mining and a scientific fascination with vacuums had come together in north-western Europe to make a rudimentary steam engine almost inevitable.

'Captain' Savery may have been a military engineer, or the rank may have been an honorary one, but he is almost as mysterious a figure as Newcomen: there is no portrait of him and the date of his birth is unknown. Like Newcomen he came from Devon. What we do know is that on 25 July 1698, the very day that Papin wrote to Leibniz about designing steam ships, Savery was granted a fourteen-year patent on 'raising water by the impellent force of fire'. The next year the patent was extended for twenty-one more years till 1733 – a rich gift to Savery's undeserving heirs, as it turned out.

Savery's machine worked as follows. A copper boiler over a fire sent steam into a water-filled tank called a receiver, where it expelled the water up a brass pipe through a non-return valve. Once the receiver was full of steam, the supply from the boiler was shut off and the receiver was sprayed with cold water, collapsing the steam inside and creating a vacuum. This sucked water up from below through a different pipe, and the cycle began again. In 1699 Savery demonstrated a version at the Royal Society with two receivers, and at some point he seems to have partly automated the mechanism of a combined valve that could fill either receiver, so the thing worked continuously.

In 1702 an advertisement said Savery's demonstration model could be inspected 'at his Workhouse in Salisbury Court, London, against the Old Playhouse, where it may be seen working on Wednesdays and Saturdays in every week from 3 to 6 in the afternoon'. He certainly sold some to the nobility, and he installed one at York Buildings, now just off the Strand but then on the banks of the Thames, where London got water from the river, but it was a failure. Mine owners were not interested. It raised water only a short distance, needed far too much coal to fuel it, leaked from its joints and blew up too easily. Failure is often the father of success in innovation.

By 1708 Papin, presumably having crossed the Channel in a conventional sailing craft rather than his own paddle boat, was in London hoping to get support to build his steam boat; we do not know if he met Savery. His hopes of being recognized as the genius of steam in England were quickly dashed. His increasingly desperate letters to Hans Sloane, Sir Isaac Newton's secretary at the Royal Society, fell on deaf ears. That he was a friend of Leibniz hardly helped. Newton's furious feud with Leibniz over who invented the calculus (they both did, but Leibniz's version was neater) was at its height, and no doubt had poisoned poor Papin's reputation by association at the Royal Society. 'There are at least six of my papers that have been read in meetings of the Royal Society and are not mentioned in the Register. Certainly, Sir, I am a sad case,' wrote Papin to Sloane in January 1712.

After that, nothing more is heard from him. He just fades away, and historians assume he must have died that year, too poor to leave a will or a record of burial. Savery would die three years later, less obscurely but hardly a national hero. He left behind one important legacy: his patent on using fire to raise water, which would force Newcomen to partner with Savery's heirs for many years.

So it is that neither of these men of science, wearing their long wigs as they mixed with grandees, managed to change the world. That was left to a humble blacksmith from Dartmouth in Devon, Thomas Newcomen. He was an ironmonger, which in those days meant something more like an engineer or blacksmith, who went into business with a glazier or plumber, John Calley, in 1685. Beyond that we know almost nothing of how he arrived at his fully fledged design of a steam engine in 1712, the year that Papin died.

Over the centuries many historians, reluctant to believe that a humble blacksmith could have succeeded where cerebral professors failed, have postulated ways in which Papin's and Savery's ideas could have reached Newcomen, including a conspiracy theory once popular in France that somebody handed Newcomen some of Papin's letters to Sloane. There is also speculation that he saw a Savery machine in a Cornish tin mine, but none of this has stood up to careful scrutiny, and it remains possible that he knew nothing of the work of the London savants. Indeed, one source insists he was at work on his first designs before 1698, the year of Savery's patent and Papin's letter to Leibniz.

That source, the only one who actually knew New-comen, was a Swede named Mårten Triewald. He worked with Newcomen and Calley, and then built several early engines in Newcastle before taking the technology back to Sweden. He describes Newcomen as experimenting with steam for a long time before getting a workable machine, and he identifies an accidental breakthrough when the injection of cold water into the cylinder was discovered:

For ten consecutive years Mr. Newcomen worked at this fire-machine which never would have exhibited the desired effect, unless Almighty God had caused a lucky incident to take place. It happened at the last attempt to make the model work that a more than wished-for effect was suddenly caused by the following strange event. The cold water, which was allowed to flow into a lead-case embracing the cylinder, pierced through an imperfection which had been mended with tin-solder. The heat of the steam caused the tin-solder to melt and thus opened a way for the cold water, which rushed into the cylinder and immediately condensed the steam, creating such a vacuum that the weight, attached to the little beam, which was supposed to represent the weight of the water in the pumps, proved to be so insufficient that the air, which pressed with a tremendous power on the piston, caused its chain to break and the piston to crush the bottom of the cylinder as well as the lid of the small boiler. The hot water which flowed everywhere thus convinced even the very senses of the onlookers that they had discovered an incomparably powerful force which had hitherto been entirely unknown in nature.

Newcomen's design collapsed the steam in a cylinder by means of this cold-water injection, and it transmitted the energy of the vacuum collapsing under the weight of the atmosphere, via a piston and a beam lever, to a pump, a mechanism safer and stronger than in Savery's design. It is probable that some full-scale versions were first built in Cornish tin mines, near where Newcomen worked, but no firm evidence has survived. The first working Newcomen engine in the world that we know of for certain was built in 1712 near Dudley Castle in Warwickshire. According to Triewald it could pump ten gallons of water twelve times a minute, lifting the water 150 feet out of the coal mine. An engraving of it by Thomas Barney in 1719 shows the beautiful complexity of the machine in sharp contrast, Rolt argues, to 'Savery's crude pump or the scientific toys of Papin'. He goes on: 'Seldom in the history of technology has so momentous an invention been developed by one man so rapidly to so developed a form.'

Yet at first it was a horribly inefficient device. A Newcomen engine is by today's standards a monster. The size of a small house, it smokes and clanks and hisses ponderously, wasting about 99 per cent of the energy in its coal fire. It would be decades before the separate condenser of James Watt, the flywheel and drive shaft, and other improvements turned it into something that could be of use in any field other than coal mining, where fuel was cheap.

I have a personal connection to this story. My ancestor, named Nicholas Ridley, got into the mining business around the end of the 1600s. Leaving a farm in the South Tyne Valley in Northumberland he became a partner in a lead-mining business and tried to smelt silver from the lead ore. He then moved to Newcastle and somehow got into

coal mining. By the time of his death in 1711 he was a prosperous coal merchant and mine owner on the north bank
of the Tyne and mayor of the town, then the third largest in
England. His son Richard ran the mines in a buccaneering
fashion, gaining a reputation as the 'stormy petrel of the
coal trade' for his propensity to get into fights and break
price-fixing cartels, even trying to murder a rival at one
point, while the second son, Nicholas, seems to have been
mostly in London, presumably receiving and marketing
the coal. Coal supplied half of England's energy as early
as 1700.

The younger Nicholas recruited the teenage Sam Calley,
son of Newcomen's partner, John, to come north and
build an engine at Byker, probably around 1715 or 1716.
This might have been the third or fourth such machine in
the world if the engineer John Smeaton is to be believed.
The Ridleys paid an enormous £400 a year in royalty to
Savery's heirs to be allowed to use this design and laid
out around £1,000 on building the first engine. This was
to drain a mine whose flooding had ruined two previous
owners.

We know this because Nicholas (junior) persuaded
Newcomen's friend Mårten Triewald to go north and oversee the youthful Calley. The Swede left an account of his
dealings with the Ridley brothers. With the success of the
first one, the Ridleys ordered more engines built and by
1733, when the Savery patent expired, there were two at
Byker, three at Heaton, one at Jesmond and one at South
Gosforth. I like to think that Richard and Nicholas Ridley
must have met Newcomen.

The Newcomen steam engine was the mother of the
modern world, ushering in an era in which technology

could begin to amplify the work of people into fantastic productivity, freeing more and more people from the drudgery of the plough, the scullery and the workhouse. It is a key innovation. Yet the way that it emerged is mysteriously obscure. Was it because of the advance of science in Britain and France, exemplified by Denis Papin? Perhaps a bit, but Newcomen apparently knew nothing of that. Was it because of improvements in metallurgy of the late seventeenth century so that large brass cylinders and pistons could now be built? Partly. Was it because of the dramatic expansion of the coal-mining industry driven by the rising price of wood as British forests shrank, and with it the demand for pumping equipment? To some extent. Was it because of the expansion of trade in north-west Europe, begun by the Dutch and leading to the creation of capital, investment and entrepreneurs? Surely yes, in part. But why did these conditions not come together in China, or Venice, or Egypt, or Bengal, or Amsterdam, or some other trading hub? And why in 1712 rather than 1612 or 1812? Innovation seems so obvious in retrospect but is impossible to predict at the time.

What Watt wrought

In 1763 a skilled and practical Scottish instrument maker, by the name of James Watt, was asked to mend a model Newcomen engine belonging to the University of Glasgow. The thing barely worked. In trying to understand what was wrong, Watt realized something about Newcomen engines in general that should have been spotted much earlier: three-quarters of the energy of the steam was being wasted in reheating the cylinder during each cycle, after it had been cooled with injected water to condense the steam. Watt had the simple idea of using a separate condenser, so that the cylinder could be kept hot, while the steam was drawn off for condensing in a cooler container. At a stroke he had improved the efficiency of the steam engine, though as usual it took months of work to get the metalworking right to make his ideas into practical devices.

After demonstrating the principle in a small test engine, Watt went into partnership with first John Roebuck to acquire a patent, then the entrepreneur Matthew Boulton to build full-scale versions. They unveiled the machine on 8 March 1776, a day before the publication of *The Wealth of Nations*, written by another Scot, Adam Smith. Boulton wanted Watt to develop a method of converting the up-and-down motion of the piston into a circular motion capable of turning a shaft for use in mills and factories. The crank and flywheel had been patented by James Pickard, which stymied Watt for a while and forced him to develop an alternative system, known as the sun-and-planets gear. Pickard in turn had got the idea of the crank from a disloyal and drunken employee of Boulton's own Soho factory, leaving the origin of this simple device mired in confusion.

Despite this example of patents getting in the way of improvement, as Savery's had for Newcomen, Watt himself was an enthusiastic defender of his own patents, and Boulton was adept at using his political contacts to acquire long-lasting and broad patents on Watt's various inventions. Just how much Watt's litigiousness delayed the expansion of steam as a source of power in factories is a hotly contested issue, but the ending of the main patent in 1800 certainly coincided with a rapid expansion of experiments and applications of steam. Indeed, one source of steady and incremental improvement in the efficiency and penetration of steam engines came as a result of the publication of a journal, *Lean's Engine Reporter*, founded by a Cornish mining engineer named John Lean, which acted like an open-software movement, disseminating suggestions for improvement among many different engineers. My point is simple: Watt, brilliant inventor though he undoubtedly was, gets too much credit, and the collaborative efforts of many different people too little.

Five years after Watt died in 1819, there was a subscription to build a monument to him, unusual in those days when monuments were mostly to those who won wars. The editors of a journal called *The Chemist* had this to say, rather perceptively: 'He is distinguished from other public benefactors, by never having made, or pretended to make it his object to benefit the public . . . This unpretending man in reality conferred more benefit on the world than all those who for centuries have made it their especial business to look after the public welfare.'

Thomas Edison and the invention business

Some time later came an energy innovation that stands symbolically for the whole field of invention: the light bulb. As a patriotic north-easterner, I cannot resist pointing out that one of the light bulb's innovators lived within a few miles of the River Tyne in Gateshead. His name was Joseph Wilson Swan. It was at the Literary and Philosophical Society in Newcastle on 3 February 1879, in front of an audience of 700 people, that he first demonstrated that he could illuminate a room – for his lecture – with an evacuated glass bulb containing a carbon filament, through which a current passed.

Electricity was already providing light by then, in the form of arc lights. The problem was that it could only be very bright. The 'subdivision' of light was the problem Swan was trying to solve, splitting a current into small flows to produce lots of sources of modest light. The realization that a glowing wire or filament did not burn up if electrified in a vacuum was critical. Creating a sufficiently empty vacuum inside blown glass and finding a material that would work reliably as a filament were the two problems Swan was trying to solve. For more than twenty years after his first prototype in 1850 he made only slow progress.

But, hang on, didn't Thomas Edison invent the light bulb? Yes, he did. But so did Marcellin Jobard in Belgium; and so did William Grove, Fredrick de Moleyns and Warren de la Rue (and Swan) in England. So too did Alexander Lodygin in Russia, Heinrich Göbel in Germany, Jean-Eugène Robert-Houdin in France, Henry Woodward and Matthew Evans in Canada, Hiram Maxim and John Starr in America, and several others. Every single one of these people produced, published or patented the idea of a glowing filament in a bulb of glass, sometimes with a vacuum, sometimes with nitrogen inside the bulb, and all before Thomas Edison.

The truth is that twenty-one different people can lay claim to have independently designed or critically improved incandescent light bulbs by the end of the 1870s, mostly independent of each other, and that is not counting those who invented critical technologies that assisted in the manufacture of light bulbs, such as the Sprengel mercury vacuum pump. Swan was the only one whose work was thorough enough and whose patents were good enough to force Edison to go into business with him. The truth is that the story of the light bulb, far from illustrating the importance of the heroic inventor, turns out to tell the opposite story: of innovation as a gradual, incremental, collective yet inescapably inevitable process. The light bulb emerged inexorably from the combined technologies of the day. It was bound to appear when it did, given the progress of other technologies.

Yet Edison, frankly, deserves his reputation, because although he may not have been the first inventor of most of the ingredients of a light bulb, and although the tale of a sudden eureka breakthrough on 22 October 1879 is largely based on retrospective mythmaking, he was none the less the first to bring everything together, to combine it with a system of generating and distributing electricity, and thereby to mount the first workable challenge to the incumbent technologies of the oil lamp and the gas lamp. So much more impressive, all told, than a blinding flash of inspiration, but vanity, vanity: people prefer to be thought brilliant rather than merely hard-working. Edison was also the one who made light bulbs (almost) reliable. Having hubristically claimed to have made a light bulb that would reliably last a long time before failing, he began a frantic search to prove his boast true. This is known today in Silicon Valley as 'fake it till you make it'. He tested more than 6,000 plant materials in his bid to try to find the ideal material for making a carbon filament. 'Somewhere in God Almighty's workshop,' Edison pleaded, 'there is a vegetable growth with geometrically powerful fibers suitable to our use.' On 2 August 1880 Japanese bamboo was the eventual winner, proving capable of lasting more than 1,000 hours.

Thomas Edison understood better than anybody before, and many since, that innovation is itself a product, the manufacturing of which is a team effort requiring trial and error. Starting his career in the telegraph industry and diversifying into stock-ticker machines, he then set up a laboratory in Menlo Park, New Jersey, in 1876, to do what he called 'the invention business', later moving to an even bigger outfit in West Orange. He assembled a team of 200 skilled craftsmen and scientists and worked them ruthlessly hard. He waged a long war against his former employee Nikola Tesla's invention of alternating-current electricity for no better reason than that Tesla had invented it rather than he. Edison's approach worked: within six years he had registered 400 patents. He remained relentlessly focused on finding out what the world needed and then inventing ways of meeting the needs, rather than the other way around. The method of invention was always trial and error. In developing the nickel-iron battery his employees undertook 50,000 experiments. He stuffed his workshops with every kind of material, tool and book. Invention, he famously said, is 1 per cent inspiration and 99 per cent perspiration. Yet in effect what he was doing was not invention, so much as innovation: turning ideas into practical, reliable and affordable reality.

And yet for all the gradual nature of the innovation of the light bulb, the result was a disruptive and trans-formational change in the way people lived. Artificial light is one of the greatest gifts of civilization, and it was the light bulb that made it cheap. A minute of work in 1880 on the average wage could earn you four minutes of light from a kerosene lamp; a minute of work in 1950 could earn you more than seven hours of light from an incandescent bulb; in 2000, 120 hours. Artificial light had come within the reach of ordinary people for the first time, banishing the gloom of winter, while expanding the opportunity to read and learn, plus incidentally reducing fire risk. There was no significant down-side to such innovation.

The incandescent bulb reigned supreme for more than a century, being still the dominant form of lighting, at least in domestic settings, well into the first decade of the twenty-first century. When it gave way to a new technology, it did so under duress. That is to say, it had to be banned, because its replacement was so unpopular. The decision by governments all over the world around 2010, lobbied by the makers of compact fluorescent bulbs, to 'phase out' incandescents by fiat in the interest of cutting carbon dioxide emissions, proved to be a foolish one. The compact fluorescent replacements took too long to warm up, did not last as long as advertised and were hazardous to dispose of. They were also much more expensive. Their energy-saving did not make up for these drawbacks in most consumers' eyes, so they had to be forced on to the market. The cost to Britain alone, of this coerced purchase and the subsidy that accompanied it, has been estimated at about £2.75bn.

Worst of all, had governments waited a few more years, they would have found a far better replacement coming along that was even more frugal in energy and had none of the disadvantages: light-emitting diodes, or LEDs. The reign of the compact fluorescents lasted just six years before they too were rapidly abandoned and manufacturers stopped producing them because of the falling cost and rising quality of LEDs. It is as if the government in 1900 had forced people to buy steam cars instead of waiting for better internal-combustion vehicles. The whole compact fluorescent light bulb episode is an object lesson in misinnovation by government. As the economist Don Boudreaux put it: 'Any legislation forcing Americans to switch from using one type of bulb to another is inevitably the product of a horrid mix of interest-group politics with reckless symbolism designed to placate an electorate that increasingly believes that the sky is falling.'

LED lights have actually been waiting in the wings for a long time. The phenomenon behind them, that semiconductors sometimes glow when conducting electricity, was first observed in 1907 in Britain and first investigated in 1927 in Russia. In 1962 a General Electric scientist named Nick Holonyak stumbled on how to make bright red LEDs of gallium arsenide phosphide, while trying to develop a new kind of laser. Yellow ones soon followed from a Monsanto lab, and by the 1980s LEDs were in watches, traffic lights and circuit boards. But until Shuji Nakamura, working for Nichia in Japan, developed a blue LED using gallium nitride in 1993, it proved impossible to make white light, which kept LED lights from mainstream lighting.

Even then it took twenty years to bring the price of this solid-state lighting down to reasonable levels. Now that has happened, however, the implications are remarkable. LED lights use so little power that a house can be well lit while not on the grid, perhaps using solar panels, a valuable opportunity for remote properties in poor countries. They have put bright flashlights inside smartphones. They emit so little heat that they make indoor 'vertical' farming of lettuces and herbs possible on a grand scale, especially using tunable LEDs to produce the wavelengths best suited to photosynthesis.

The ubiquitous turbine

If Newcomen was from humble origins, poor and illiterate in his younger days, the same cannot be said of another key name in the story of steam. Charles Parsons was the sixth son of the wealthy Earl of Rosse, an Irish peer. He was born and raised at Birr Castle in County Offaly, Ireland, and given private tuition in place of school before going up to Cambridge University to read mathematics.

But this was no typical aristocratic household. The earl was an astronomer and engineer. He encouraged his sons to spend time in his workshops rather than libraries. Charles and his brother built a steam engine with which to provide the power for grinding the reflector on his father's telescope. When he left university it was not for a comfortable berth in the law, politics or finance, but for an apprenticeship in an engineering firm on the Tyne. He proved a brilliant engineer and in 1884 he designed and patented the steam turbine that would prove to be, with very few modifications, the indispensable machine that gave the world electricity and that powered the navies and liners of the sea and later the jets of the air. To this day, it is basically Parsons's design that keeps the lights on, navies afloat and airliners aloft.

A turbine is a device that spins on its axis. There are two ways to use steam (or water) to make something turn: impulse or reaction. Directing the steam from a fixed nozzle at buckets on a wheel will turn that wheel; and squirting the steam at an angle out of nozzles on the outsides of a wheel itself will also turn the wheel. A spinning sphere driven by steam shooting out of two angled nozzles had been built as a toy by Hero of Alexandria in the first century AD. Parsons concluded early on that impulse turbines were

inefficient and stressful to the metal. He realized too that a
series of turbines, each turned by some of the steam, would
gather more of the energy more efficiently. He redesigned
dynamos to generate electricity from turbines and within a
few years the first electric grids were being built with larger
and larger Parsons turbines.

Parsons set up his own company but had to leave behind
the intellectual property in his original designs, and he spent
five years trying to build radial-flow turbines before he was
able to revert to parallel-axial-flow turbines. He tried and
failed to interest the Admiralty in the devices as a way of
powering ships. So in 1897 he sprang a cheeky surprise on
the Royal Navy.

Parsons, who was fond of boats and yachting, had made
a sleek little ship, *Turbinia*, powered by steam turbines turning a screw propeller. The first results were disappointing,
mainly because of the propeller, which caused 'cavitation'
in the water – small vacuum pockets behind the screw
blades that wasted energy. Parsons and Christopher Leyland went back to the laboratory, trying many designs to
find one that might solve the cavitation problem. It was
trial and error. They stayed up all night at times and were
still at the water tank when the housemaids arrived in the
morning. It was frustrating work, but by 1897 Parsons had
replaced the single radial-flow turbine with three axial-flow
ones, and the single propeller shaft with three shafts, each
armed with three screws. He knew by now, from sea trials,
that his little craft, with nine propellers, could achieve
34 knots, much faster than any ship of the time. He even
gave a public talk about it in April 1897, which the *Times*
newspaper reported, concluding dismissively that turbine
technology was 'in a purely experimental, perhaps almost

in an embryo stage' as far as ships were concerned. How wrong they were.

As the Grand Fleet assembled at Spithead on 26 June in the presence of the Prince of Wales, to mark the Diamond Jubilee of Queen Victoria, Parsons was planning an audacious stunt. Over 140 ships were drawn up in four lines over twenty-five miles long in all. Between them steamed a royal procession of ships: *Victoria and Albert*, carrying the Prince of Wales, the P&O liner *Carthage*, with other royal guests aboard, *Enchantress*, with the Lords of the Admiralty, *Danube*, with members of the House of Lords, *Wildfire*, with colonial prime ministers, the Cunard liner *Campania*, with members of the House of Commons, and finally *Eldorado*, carrying foreign ambassadors. A line of invited foreign battleships included the *König Wilhelm* with Prince Henry of Prussia aboard.

Defying the rules and evading the fast steam boats on picket duty, Parsons took *Turbinia* between the ranks of battleships at full speed and then steamed up and down in front of the grandees, pursued in vain by Royal Navy vessels, one of which almost collided with the little greyhound of the sea. It was a sensation. With surprisingly little umbrage – it helped that the Germans were there to witness the episode, and Prince Henry of Prussia took care to send a congratulatory message to Parsons – the Navy took the hint and by 1905 had determined that all future warships would be turbine-powered. HMS *Dreadnought* was the first. In 1907, the vast liner *Mauretania*, powered by Parsons turbines, was photographed alongside her little predecessor, *Turbinia*.

The Spithead moment is in some ways misleading. The history of turbines and electricity is profoundly gradual, not marked by any sudden step changes. Parsons was just one of many people along the path who incrementally devised and improved the machines that made electricity and power. It was an evolution, not a series of revolutions. The key inventions along the way each built upon the previous one and made the next one possible. Alessandro Volta made the first battery in 1800; Humphry Davy made the first arc lamp in 1808; Hans Christian Oersted made the connection between electricity and magnetism in 1820; Michael Faraday and Joseph Henry made the first electric motor in 1820 and its opposite, the first generator, in 1831. Hippolyte Pixii made the first dynamo in 1832; Samuel Varley, Werner von Siemens and Charles Wheatstone all came up with the full dynamo-electric generator in 1867; Zénobe Gramme turned this into a direct-current generator in 1870.

Parsons's turbine was about 2 per cent efficient at turning the energy of a coal fire into electricity. Today a modern combined-cycle gas turbine is about 60 per cent efficient. A graph of the progress between the two shows a steady improvement with no step changes. By 1910, using waste heat to preheat the water and the air, engineers had improved the efficiency to 15 per cent. By 1940, with pulverized coal, steam reheating and higher temperatures, it was nearer 30 per cent. In the 1960s, as the combined-cycle generator effectively brought a version of the turbojet engine in alongside the steam turbine, potential efficiency had almost doubled again. To single out clever people who made the difference along the way is both difficult and misleading. This was a collaborative effort of many brains. Long after the key technologies had been 'invented', innovation continued.

Nuclear power and the phenomenon of disinnovation

The twentieth century saw only one innovative source of energy on any scale: nuclear power. (Wind and solar, though much improved and with a promising future, still supply less than 2 per cent of global energy.) In terms of its energy density, nuclear is without equal: an object the size of a suitcase, suitably plumbed in, can power a town or an aircraft carrier almost indefinitely. The development of civil nuclear power was a triumph of applied science, the trail leading from the discovery of nuclear fission and the chain reaction through the Manhattan Project's conversion of a theory into a bomb, to the gradual engineering of a controlled nuclear fission reaction and its application to boiling water. No individual stands out in such a story unless it be Leo Szilard's early realization of the potential of a chain reaction in 1933, General Leslie Groves's leadership of the Manhattan Project in the 1940s, or Admiral Hyman Rickover's development of the first nuclear reactors and their adaptation to submarines and aircraft carriers in the 1950s. But as these names illustrate, it was a team effort within the military and state-owned enterprises, plus private contractors, and by the 1960s it had culminated in a huge programme of constructing plants that would use small amounts of enriched uranium to boil enormous amounts of water reliably, continuously and safely all over the world.

Yet today the picture is of an industry in decline, its electrical output shrinking as old plants close faster than new ones open, and an innovation whose time has passed, or a technology that has stalled. This is not for lack of ideas, but for a very different reason: lack of opportunity to experiment. The story of nuclear power is a cautionary tale of how innovation falters, and even goes backwards, if it cannot evolve.

The problem is cost inflation. Nuclear plants have seen their costs relentlessly rising for decades, mostly because of increasing caution about safety. And the industry remains insulated almost entirely from the one known human process that reliably pulls down costs: trial and error. Because error could be so cataclysmic in the case of nuclear power, and because trials are so gigantically costly, nuclear power cannot get trial and error restarted. So we are stuck with an immature and inefficient version of the technology, the pressurized-water reactor, and that is gradually being strangled by the requirements of regulators acting on behalf of worried people reacting to anti-nuclear activists. Also, technologies pushed on the world by governments, before they are really ready, sometimes falter, where they might have done better if allowed to progress a little more slowly. The transcontinental railroads in the United States were all failures, resulting in bankruptcies, except the one privately funded one. One cannot help thinking that nuclear power developed in less of a hurry, and less as a result of a military spin-off, might have done better.

In a book published in 1990, *The Nuclear Energy Option*, the nuclear physicist Bernard Cohen argued that the reason we stopped building nuclear plants in the 1980s in most of the West was not from fear of accidents, leaks or the proliferation of atomic waste; it was instead the inexorable escalation of costs driven by regulation. His diagnosis has proved even more true since.

This is not for want of ideas for new kinds of nuclear

power. There are hundreds of different designs for fission reactors out there in engineers' PowerPoint presentations, some of which have reached working-prototype design in the past and would have gone further if offered as much financial support as the conventional light-water reactor. Liquid-metal and liquid-salt reactors are two broad categories. The latter would work using salts of thorium or uranium fluoride, probably with other elements included such as lithium, beryllium, zirconium or sodium. The key advantage of such a design is that the fuel comes in liquid form, rather than as a solid rod, so cooling is more even and the removal of waste easier. There is no need to operate at high pressure, reducing the risks. The molten salt is the coolant as well as the fuel and has the neat property that the reaction slows down as it gets hotter, making meltdown impossible. In addition, the design would include a plug that would melt above a certain temperature, draining the fuel into a chamber where it would cease fission, a second safety system. Compared with, say, Chernobyl, this is dramatically safer.

Thorium is more abundant than uranium; it can in effect breed almost indefinitely by creating uranium 233; it can generate almost 100 times as much power from the same quantity of fuel; it does not give rise to fissile plutonium; it generates less waste with a shorter half-life. But although a submarine with sodium coolant was launched in the 1950s and two experimental thorium molten-salt reactors were built in the 1960s in the United States, the project eventually expired as all the money, training and interest focused on the light-water uranium design. Various countries are looking at how to reverse this decision, but none has really taken the plunge.

Even if they did, it seems unlikely that they would achieve the notorious promise made in the 1960s that nuclear power would one day be 'too cheap to meter'. The problem is simply that nuclear power is a technology ill-suited to the most critical of innovation practices: learning by doing. Because each power station is so big and expensive, it has proved impossible to drive down the cost by experiment. Even changing the design halfway through construction is impossible because of the immense regulatory thicket that each design must pass through before construction. You must design the thing in advance and stick to that design or go back to square one. This way of doing things would fail to bring down costs and raise performance in any technology. It would leave computer chips at the 1960 stage. We build nuclear power stations like Egyptian pyramids, as one-off projects.

Following the Three-Mile Island accident in 1979, and Chernobyl in 1986, activists and the public demanded greater safety standards. They got them. According to one estimate, per unit of power, coal kills nearly 2,000 times as many people as nuclear; bioenergy fifty times; gas forty times; hydro fifteen times; solar five times (people fall off roofs installing panels) and even wind power kills nearly twice as many as nuclear. These numbers include the accidents at Chernobyl and Fukushima. Extra safety requirements have simply turned nuclear power from a very, very safe system into a very, very, very safe system.

Or maybe they have made it less safe. Consider the Fukushima disaster of 2011. The design at Fukushima had huge safety flaws. Its pumps were in a basement easily flooded by a tidal wave, a simple design mistake unlikely to be repeated in a more modern design. It was an old reactor and would have been phased out long since if Japan had still been building new nuclear reactors. The stifling of nuclear expansion and innovation through costly over-regulation had kept Fukushima open past its due date, thus lowering the safety of the system.

The extra safety demanded by regulators has come at high cost. The labour that goes into the construction of a nuclear plant has hugely increased, but mostly in the white-collar jobs, signing off paperwork. According to one study, during the 1970s new regulations increased the quantity of steel per megawatt by 41 per cent, concrete by 27 per cent, piping by 50 per cent and electrical cable by 36 per cent. Indeed, as the ratchet of regulation turned, the projects began to add features to anticipate rule changes that sometimes did not even happen. Crucially this regulatory environment forced the builders of nuclear plants to drop the practice of on-the-spot innovation to solve unanticipated problems, lest it lead to regulatory resets, which further drove up cost.

The answer, of course, is to make nuclear power into a modular system, with small, factory-built reactor units produced off production lines in large quantities and installed like eggs in a crate at the site of each power station. This would drive down costs as it did for the Model T Ford. The problem is that it takes three years to certify a new reactor design, and there is little or no short-cut for a smaller one, so the cost of certification falls more heavily on a smaller design.

Meanwhile, it is now likely that nuclear fusion, the process of releasing energy from the fusion of hydrogen atoms to form helium atoms, may at last fulfil its promise and begin to provide almost unlimited energy within the next few decades. The discovery of so-called high-temperature superconductors and the design of so-called spherical toka-maks have probably at last defused the old joke that fusion power is thirty years away – and has been for thirty years. Fusion may now come to commercial fruition, in the form of many relatively small reactors generating electricity, maybe 400 megawatts each. It is a technology that brings almost no risk of explosion or meltdown, very little in the way of radioactive waste and no worries about providing material for weapons. Its fuel is mainly hydrogen, which it can make with its own electricity from water, so its foot-print on the earth will be small. The main problem fusion will still have to solve, as with nuclear fission, is how to drive down the cost by mass production of the reactors, with the ability to redesign from experience along the way so as to learn cost-cutting lessons.

Shale gas surprise

One of the most surprising stories of the twenty-first century has been the rise of natural gas, a fuel that just a decade ago was thought to be on the brink of running out and is now both cheap and plentiful. It is mainly the story of the innov-ation that led to the production of gas from shale. Right up till 2008 or so, it was conventional wisdom among energy experts that cheap natural-gas supplies would be exhausted to all practical extent fairly early in the twenty-first cen-tury. Oil and coal would last longer. This prediction had been made before, repeatedly. In 1922 the US Coal Com-mission, set up by President Warren Harding, interviewed 500 people in the energy industry over eleven months, and came to the conclusion that 'already the output of gas has begun to wane'. In 1956 the oil expert M. King Hubbert predicted that natural-gas production in the United States would peak in 1970 at 38 billion cubic feet per day and decline. In fact it was 58 bcf a day then and still rising. Today it is over 80 bcf per day.

These predictions proved gloriously wrong, for two reasons. First, in America, strict price regulation of gas in the 1970s, based on the theory that it was scarce, effectively halted gas exploration in its tracks. Companies flared off or shut down gas as a nuisance, and pursued oil instead. This did indeed produce a peak in production which many mistook for the beginning of exhaustion of reserves. Incredibly, the US government passed several measures in the 1970s to forbid the generation of electricity by oil or gas in any utility that could get access to coal, and forbade the building of plants that could not use coal. Deregulation of the gas industry under President Reagan led to a surge in production.

The second reason for the gas glut of the second decade of the twenty-first century was innovation. Throughout the United States, gas and oil exploration companies set out to find ways to squeeze more out of each field, and to squeeze gas and oil out of 'tight' rocks, whence it did not flow naturally. This resulted in the serendipitous discovery of 'slick-water' hydraulic fracturing in the 1990s in Texas, which, combined with the new ability to drill round corners, and thus go horizontally within seams of rock for miles on end, made tight shales, where most hydrocarbons are stored, into huge sources of gas and oil. Add in offshore gas, plus the ability to liquefy gas for transport by sea, and it becomes clear why the world now has ample supplies of gas, the cleanest, lowest-carbon and safest of the fossil fuels.

The key location of the slick-water fracking breakthrough was the Barnett shale near Fort Worth, where
an entrepreneur named George Mitchell, born to a Greek
goatherd father, had grown rich supplying Chicago with
gas. He had a good fixed-price contract. If he moved elsewhere he would have to drop his price. So he was desperate
to squeeze more from the Barnett shale, where he had lots
of drilling rights. By the late 1990s output was dropping,
and so was Mitchell Energy's share price, which was causing Mitchell personal difficulties, because of commitments
he had made to philanthropy, backed by loans against his
shares. His wife had Alzheimer's and he had prostate problems. By rights the 78-year-old multimillionaire should
have been reasonable, should have given up on America
as the oil majors were already doing, and cut his losses.
The future of gas lay offshore, or in Russia and Qatar. But
Mitchell, like many innovators, was not reasonable, so he
kept trying to get the gas to flow.

The Barnett shale was known to be rich in hydrocarbons,
but they would not flow easily, so the rock needed to be
cracked deep underground, and the microscopic cracks
propped open. A technology to do this was well known,
and relied on gels to prop open the cracks and let the gas
out. It worked well in some rocks but not in shale. Mitchell
sank $250m into trying to make it work in the Barnett field
without success.

One day in 1996 a Mitchell employee named Nick
Steinsberger noticed an odd result. He was employing contractors to pump a stiff gel with large amounts of sand in
it down the well. But since gel and sand were expensive, he
had been forcing the service companies to lower the amount
of gel and chemicals in the mixture pumped down the hole

in an attempt to lower costs and pump less of the viscous material into the shale. On this day, the gel was so dilute it would not 'gel' properly. Steinsberger pumped it down the hole anyway and noticed the well produced a decent surge of gas. He tried some more wells with similar results. Attending a baseball game with a friend from another company, Mike Mayerhofer, he heard a similar story – water with a little lubricant and much less sand was working well in a different kind of rock, in this case tight sandstone in east Texas.

So in 1997 Steinsberger then began deliberately using a more watery liquid, basically water mixed with less sand and a very small quantity of ordinary kitchen sink chemicals (bleach and soap, essentially), instead of gel. He tried this on three wells, but it did not work. 'The pressure went up too high, forcing me to terminate the pump job, because the slickwater wouldn't carry the sand in shale like it would in much more permeable tight sands.' In early 1998, getting pretty desperate and with his bosses ready to give up on the Barnett shale, he convinced management to let him try three more wells. This time he pumped a lot more slick water but increased the sand from extremely low concentrations to higher over the course of the job. The first well, S. H. Griffin Estate 4, produced a surge of gas and kept on doing so for weeks and months. He realized he had stumbled on a formula that was not just half as expensive, but twice as productive. A flash in the pan? No, the other two wells had similar results.

Steinsberger's breakthrough transformed the last years of George Mitchell's life, turning him into a billionaire when he sold his company. It turned the Barnett shale into America's largest gas producer. Copied elsewhere, and steadily improved by further innovation, it had the same effect in shale after shale, in Louisiana, Pennsylvania, Arkansas, North Dakota, Colorado, then Texas again. Soon the same technique was being adapted to get oil out as well. Today America is not only the world's biggest producer of gas; it is also the world's biggest producer of crude oil, thanks entirely to the shale-fracking revolution. The Permian basin in Texas alone now produces as much oil as the whole of the United States did in 2008, and more than any OPEC country except Iran and Saudi Arabia. America was building huge gas import terminals in the early 2000s; these have now been converted into export terminals. Cheap gas has displaced coal in the country's electricity sector, reducing its emissions faster than any other country. It has undermined OPEC and Russia, leaving the latter frantically supporting anti-fracking activists to try to defend its markets – with much success in innovation-phobic Europe, where shale exploitation has been largely prevented.

A cheap-gas, cheap-oil glut brought on deliberately by OPEC in 2015 to try to bust the frackers had the opposite effect, killing weaker companies but forcing the survivors to work out how to remain competitive at sixty, fifty and forty dollars per barrel of oil. The availability of cheap hydrocarbons gave American manufacturing an edge, resulting in a rapid 'reshoring' of chemical industries to the United States and a surge of chemical companies leaving Europe. The energy policies of a dozen countries like Britain, predicated on ever-rising fossil-fuel energy prices to make wind and nuclear look less expensive, became expensive follies almost overnight.

Why did this revolution happen in America, an old, played-out and well-explored oil and gas region? The answer lies partly in property rights. Because of mineral rights belonging to local landowners, rather than the state, and because oil companies had never been nationalized, as they were in so many other countries, from Mexico to Iran, America had a competitive, pluralistic and entrepreneurial oil-drilling mindset, manifested in a 'wild-cat' industry, backed by deep pockets of risk capital – the early frackers spent vast sums of borrowed money before turning cash-positive. As one account of the story by the key innovators put it:

> Small companies often have the upper hand in leasing mineral rights from landowners as their interaction with landowners is generally more personalized. Shale production was hotly pursued by many small companies resulting in a multitude of varied drilling and completion methods being implemented and tested across multiple basins. These 'laboratories' have resulted in continuous improvements and fostered economic success.

So trial and error was vital to innovation in fracking. Steinsberger made a series of lucky mistakes, failing many times along the way. And when he had found the formula, he did not know why it worked. A seismology expert, Chris Wright, soon explained it. Wright, an engineer whose company, Pinnacle, was using new tiltmeter devices to help track the progress of fractures underground for Mitchell, figured out that slick-water fracs created large networks of multiple fractures. He had developed a model of simultaneous growth of multiple fractures in the early 1990s 'which was widely derided by all the old-timers in the frac world as they insisted multiple fracs would always rapidly coalesce into a single frac'. It turned out Wright was right. The pressurized water was creating cross-cutting fractures in the rocks, greatly increasing the surface area exposed to the sand. Fractures were propagating a mile or more in one direction, but spreading hundreds of metres either side of this axis too. In this case science came in behind the technology, rather than vice versa. Recent attempts to credit the federal government with starting this innovation mostly miss the point. Yes, lots of research was done at government laboratories, but much of it under contract to the gas industry, and largely because there were entrepreneurs like Mitchell and Wright (now one of the industry leaders) creating the demand for such research.

At first environmentalists welcomed the shale gas revolution. In 2011 Senator Tim Wirth and John Podesta welcomed gas as 'the cleanest fossil fuel', writing that fracking 'creates an unprecedented opportunity to use gas as a bridge fuel to a 21st-century energy economy that relies on efficiency, renewable sources, and low-carbon fossil fuels such as natural gas'. Robert Kennedy, Jr, head of the Waterkeeper Alliance, wrote in the *Financial Times* that 'In the short term, natural gas is an obvious bridge fuel to the "new" energy economy.' But then it became clear that this cheap gas would mean the bridge was long, posing a threat to the viability of the renewable-energy industry. Self-interest demanded a retraction by Kennedy, which he duly provided, calling shale gas a 'catastrophe'.

In the heartlands where fracking began, Texas, Louisiana, Arkansas and North Dakota, there was little opposition. A lot of empty land, a long tradition of oil drilling and a culture of can-do enterprise ensured that the shale revolution prospered unhindered by much if any local protest. But when it spread to the East Coast, to Pennsylvania and then New York, suddenly shale gas began to attract enemies, and environmentalists spotted an opportunity to fund-raise on the back of opposition. Recruiting some high-profile stars, including Hollywood actors such as Mark Ruffalo and Matt Damon, the bandwagon gathered pace. Accusations of poisoned water supplies, leaking pipes, contaminated waste water, radioactivity, earthquakes and extra traffic multiplied. Just as the early opponents of the railways accused trains of causing horses to abort their foals, so no charge was too absurd to level against the shale gas industry. As each scare was knocked on the head, a new one was raised. Yet despite millions of 'frac jobs' in thousands of wells, there were very few and minor environmental or health problems.

The reign of fire

One of the flaws in the way we recount stories of innovation is that we unfairly single out individuals, ignoring the contribution of lesser mortals. I have chosen to tell the stories of Newcomen, Watt, Edison, Swan, Parsons and Steinsberger, but they were all stones in an arch or links in a chain. And not all of them ended up wealthy, let alone their descendants. There is no foundation named after any of them today and funded by their wealth. It was the rest of us who reaped most of the benefit of their innovations.

Yet energy itself does deserve to be singled out. It is the root of all innovation if only because innovation is change and change requires energy. Energy transitions are crucial, difficult and slow. For the vast majority of history, argues John Constable, the supply of energy, from wheat and wind

and water, was just too thin to generate complex structures on a sufficient scale to transform people's lives. Along
came the heat-to-work transition of 1700 and suddenly it
became possible to create ever more improbable and complex material structures from the harnessing of fossil fuels
with their huge energy yield on energy invested. The fossil-
fuel dependence of the modern world is roughly the same
today – at about 85 per cent of primary energy – as it was
twenty years ago. The vast majority of society's need for
energy is supplied by heat. What will eventually depose the
'impellent use of fire', that strange link between heat and
work that came into the lives of humanity around the year
1700 and is still vital to the world? Nobody yet knows.

2

Public health

An operation invented not by persons conversant in philosophy or skilled in physic, but by a vulgar, illiterate people; an operation in the highest degree beneficial to the human race.

GIACOMO PYLARINI on smallpox inoculation, 1701

Lady Mary's dangerous obsession

In the same year that Thomas Newcomen was building his first steam engine, 1712, and not far away, a more romantic episode was in train, and one that would indirectly save even more lives. It was much higher up the social scale. Lady Mary Pierrepoint, a well-read, headstrong young woman of twenty-three, was preparing to elope in order to escape the prospect of a dull marriage. Her wealthy suitor, Edward Wortley Montagu, with whom she had carried on a voluminous correspondence characterized by furious disagreement as well as outrageous flirtation, had failed to agree a marriage settlement with her even wealthier father, the Earl (later Duke) of Kingston. But the prospect of being forced by her father to marry instead a pecunious dullard, the Honourable Clotworthy Skeffington, persuaded Mary to rekindle the romance with Wortley (as she called him). She proposed elopement, and he, despite thus missing out on her dowry, and in a fit of uncharacteristic impetuosity, agreed. The episode turned to farce: he was late, she set off for the rendezvous alone, he overtook her at an inn but did not realize she was there, but after further mishaps they found each other and married on 15 October 1712 in Salisbury.

After this romantic start the marriage was a disappointment, Wortley proving a cold and unimaginative husband. His bride – learned, eloquent and witty – cut a swathe through literary London, writing eclogues with Alexander Pope in the style of Virgil, and befriending the literary lions and social tigers of the day. Joseph Spence would later write: 'Lady Mary is one of the most extraordinary shining characters in the world; but she shines like a comet; she is all irregular and always wandering. She is the most wise, most imprudent; loveliest, disagreeablest; best natured, cruellest woman in the world.'

Then smallpox marked her skin and made her reputation. This vicious virus, humankind's greatest killer, was constantly a threat in early-eighteenth-century London. It had recently killed Queen Mary and her nephew, the young Duke of Gloucester, the last Stuart heir to the throne who was not Catholic; it had almost killed the Electress of Hanover, Sophia, and her son George, destined to be the next king of England instead. It killed Lady Mary's brother in 1714 and very nearly killed her the next year, leaving her badly scarred and lacking in eyelashes, her beauty cruelly ravaged.

But it was smallpox that would bring her lasting fame, for she became one of the first, and certainly one of the most passionate, champions in the Western world of the innovative practice of inoculation. In 1716 her husband was sent as ambassador to Constantinople and Lady Mary accompanied him with her young son. She did not invent inoculation, she did not even bring the news of it for the first time, but being a woman she was able to witness in detail the practice among women cloistered in Ottoman society, and then to champion it back home among mothers terrified for their children, to the point where it caught on. She was an innovator, not an inventor.

Two reports had reached the Royal Society in London from Constantinople of the practice of 'engrafting' as a cure for smallpox. According to the correspondents, Emmanuel Timonius and Giacomo Pylarini, both physicians working in the Ottoman Empire, the pus from a smallpox survivor would be mixed with the blood in a scratch on the arm of a healthy person. The reports were published by the Royal Society but dismissed as dangerous superstition by all the experts in London. More likely to spark an epidemic than prevent it; an unconscionable risk to be running with people's health; an old wives' tale; witchcraft. Given the barbaric and unhelpful practices of doctors at the time, such as bloodletting, this was both ironic and perhaps understandable.

It seems the Royal Society had been told of the practice even earlier, in 1700, by two correspondents in China, Martin Lister and Clopton Havers. So there was nothing new about this news. But where these doctors failed to persuade the British, Lady Mary Wortley Montagu had better luck. On 1 April 1718 she wrote to her friend

Sarah Chiswell from Turkey with a detailed account of inoculation:

> The smallpox, so fatal and so general amongst us, is here entirely harmless by the invention of engrafting, which is the term they give it. There is a set of women who make it their business to perform the operation . . . When they are met (commonly fifteen or sixteen together) the old woman comes with a nutshell full of the matter of the best sort of smallpox, and asks what veins you please to have opened. She immediately rips open that you offer to her with a large needle (which gives you no more pain than a common scratch) and puts into the vein as much venom as can lie upon the head of her needle . . . There is no example of any one that has died in it, and you may believe I am well satisfied of the safety of the experiment, since I intend to try it on my dear little son. I am patriot enough to take pains to bring this useful invention into fashion in England.

Lady Mary did indeed engraft her son Edward, anxiously watching his skin erupt in self-inflicted pustules before subsiding into immunized health. It was a brave moment. On her return to London she inoculated her daughter as well, and became infamous for her championing of the somewhat reckless procedure – a sort of version of the trolley problem so beloved of moral philosophers: do you divert a runaway truck from a line where it will kill five people to another line where it will kill one? Do you deliberately take one risk to avoid a greater one? By then, some doctors had joined the cause, notably Charles Maitland. His inoculation of the children of the Prince of Wales in 1722 was a significant moment in the campaign. But even afterwards there was furious denunciation of the barbaric practice. Misogyny and prejudice lay behind some of it, as when Dr William Wagstaffe pronounced: 'Posterity will scarcely be brought to believe that an experiment practised only by a few ignorant women amongst an illiterate and unthinking people should on a sudden – and upon a slender Experience – so far obtain in one of the politest nations in the world as to be received into the Royal Palace.'

In America, the practice of inoculation arrived around the same time, through the testimony of an African slave named Onesimus, who told the Boston preacher Cotton Mather about it, possibly as early as 1706, who in turn informed the physician Zabdiel Boylston. For trying inoculation on 300 people, Boylston was subject to fierce criticism and life-threatening violence, abetted by rival physicians, to the point where he had to hide for fourteen days in a secret closet lest the mob kill him. Innovation often requires courage.

In due course inoculation with smallpox itself – later known as variolation – was replaced by the safer but similar practice of vaccination, that is to say, using a related but less dangerous virus than smallpox, an innovation usually credited to Edward Jenner. In 1796 he deliberately infected an eight-year-old boy, James Phipps, with cowpox from blisters on the hands of a milkmaid called Sarah Nelmes, who had caught it from a cow called Blossom. He then tried to infect Phipps with smallpox itself and showed that he was immune to it. This demonstration proof, not the vaccination itself, was his real contribution and the reason he had such an impact. The idea of deliberately giving people cowpox to immunize them against smallpox was by then already thirty years old. It had been tried by a physician named John Fewster in 1768, and by several other doctors in Germany and England in the 1770s. It was already probably in use among farmers before that date.

So, yet again, innovation proves to be gradual and to begin with the unlettered and ordinary people, before the elite takes the credit. That is perhaps a little unfair on Jenner, who, like Lady Mary Wortley, deserves fame for persuading the world to adopt the practice. Napoleon, despite being at war with Britain, had his armies vaccinated, on the strength of Jenner's advocacy, and awarded Jenner a medal, calling him 'one of the greatest benefactors of mankind'.

Pasteur's chickens

Vaccination conquered smallpox so comprehensively that by the 1970s the disease – once the greatest taker of human lives on the planet – had died out altogether. The last case of the more deadly strain, *Variola major*, was in Bangladesh in October 1975. Rahima Banu, then three years old, survived and is still alive. The last case of *Variola minor* was in October 1977 in Somalia. Ali Maow Maalin, who was an adult when he caught it, also survived, working for most of his life on the campaign against polio, and dying in 2013 of malaria.

Vaccination exemplifies a common feature of innovation: that use often precedes understanding. Throughout history, technologies and inventions have been deployed successfully without scientific understanding of why they work. To a rational person in the eighteenth century, Lady Mary's idea that exposure to one strain of a fatal disease could protect against that disease must have seemed crazy.

There was no rational basis to it. It was not until the late nineteenth century that Louis Pasteur began to explain how and why vaccination worked.

Pasteur proved that germs were microscopic organisms by boiling a fermented liquid and showing that it remained inert and could not generate further fermentation unless exposed to germs carried in on the breeze. His final blow was to leave the liquid open to the air but only through a narrow swan-necked vessel whose shape ensured that bacteria did not pass through. He boasted, in 1862: 'Never will the doctrine of spontaneous generation recover from the mortal blow of this simple experiment.'

If contagious diseases were caused by microbes – the distinction between bacteria and far smaller viruses was still to be made – then could inoculation be explained by a change in the character of the microbe and a change in the human body's vulnerability to it? Pasteur's explanation came about as a result of a serendipitous accident. In the summer of 1879 he went on holiday, leaving his assistant, Charles Chamberland, to inoculate some chickens with cholera from an infected chicken broth, as part of a series of experiments to understand the nature of the cholera bacterium. Chamberland forgot and went on holiday himself. When they returned from holiday, and did the experiment, the stale broth proved capable of making the chickens ill, but not killing them.

Acting perhaps on a hunch, Pasteur now turned to a virulent cholera strain that normally killed chickens easily and injected it into these now recovered (and long-suffering) birds. It failed even to sicken them, let alone kill them. The weak strain of cholera had immunized them against this stronger strain. Pasteur began to realize that vaccination worked by a less virulent organism triggering an immune response that worked against a more virulent one. Not that he understood yet the slightest thing about the human immune system. Science was beginning to catch up with technology.

The chlorine gamble that paid off

The scene is a court house in New Jersey and the year is 1908. On trial is the Jersey City Water Supply Company, which had lost a previous case in which it had been proved that the company was not supplying 'pure and wholesome water' to the city, as specified by its contract. The problem was that upstream of the city's reservoir more and more people were building homes and discharging sewage from privies directly into the streams that fed the reservoir. Deaths from typhoid were far too common in the city. Despite removing more than 500 such privies since 1899 and filtering the water, the company could not prevent the contamination of the water supply happening two or three times a year after heavy rain.

Given three months to rectify the situation by the court, the company's sanitary adviser, Dr John Leal, came up with the idea of dripping chloride of lime, a disinfectant, into the water supply. By 26 September, three days before the second trial began, the plant was built, operating and continually chlorinating 40 million gallons of water a day. It emerged during the trial that Leal had sought nobody's permission to conduct this experiment on the citizens of Jersey City, at a time when there was widespread revulsion at the idea of putting chemicals into drinking water. 'The idea itself of chemical disinfection is repellent,' thundered the suitably named Thomas Drown of the Massachusetts Institute of Technology, echoing the views of others in the establishment. Dr Leal's decision was brave and risky.

In court, the city's lawyer therefore objected to the notion that the company could meet its responsibility to provide clean water with a quick chemical fix of unknown risk to the population – which had not been asked for its consent. He requested the judge refuse even to hear the evidence about whether chlorination had helped. The judge disagreed and allowed the company to present its case. Under cross-examination, Leal said of chlorination: 'I think that it is the safest, the easiest, and the cheapest and best method for rendering this water pure every day in the year and every minute of every hour.' He added: 'I believe the water supply of Jersey City today is the safest in the world.'

Q. Any ill effect on the health of the people there?
A. Not the slightest.
Q. Do you drink this water?
A. Yes, sir.
Q. Habitually?
A. Yes, sir.

After a lengthy trial the judge eventually ruled that the company had met its responsibilities by this innovation. The Jersey City case proved a turning point, a clean-water-shed. Cities all over the country and the world began using chlorination to clean up water supplies, as they do to this day. Typhoid, cholera and diarrhoea epidemics rapidly disappeared. But where did Dr Leal get the idea? From a similar experiment in Lincoln in England, he said at the trial. Like most innovators he did not claim to be the inventor.

The city of Lincoln had seen death rates from typhoid decline after the installation of a sand-filtration plant for its water supply. But in 1905 it was struck by a bad outbreak: 125 people died. The city called in Dr Alexander Cruickshank Houston, a bacteriologist with the Royal Commission on Sewage Disposal. Within two days of arriving in February 1905, Dr Houston had jerry-rigged a device for dripping Chloros (sodium hypochlorite) by gravity into the water, with immediate results on the rate of new typhoid infections.

But where did Dr Houston get the idea? Perhaps from an Indian Army Medical Services officer by the name of Vincent Nesfield, who published a paper in 1903 suggesting exactly how to make and use liquid chlorine to disinfect water supplies. Nesfield's technique is close to what is used today, and well ahead of its time. Whether and where he ever used it is unknown.

And where did Dr Nesfield get the idea? Perhaps from a typhoid outbreak in Maidstone in Kent in the autumn of 1897, in which 1,900 people caught the disease and about 150 died. Here, 'under the supervision of Dr Sims Woodhead, acting on behalf of the water company, the reservoir and mains of the Farleigh area of water supply at Maidstone were on Saturday night disinfected with a solution of chloride of lime.' By December the outbreak was over.

And where did Dr Woodhead get the idea? Probably from the use of chloride of lime as a disinfectant of sewage, which was by then a well-known technique. Chloride of lime had also by this time caught on as an antiseptic among surgeons, though they were disgracefully slow as a profession to realize that they should be washing their hands at all let alone with strong bleach.

During London's cholera epidemic of 1854, chloride of lime was used so liberally in Soho that, as one magazine reported, 'The puddles are white and milky with it, the stones are smeared with it; great splashes of it lie about in the gutters, and the air is redolent with its strong and not very agreeable odour.'

At the time of that London epidemic, Dr John Snow was trying, largely in vain, to persuade the authorities that cholera was caused by dirty water, not smelly air – the 'miasma' theory then in vogue. He had shown that those getting their water supplies from the Thames estuary were far more likely to catch cholera than those getting their supply from rural streams, and he famously removed the handle from a water pump in Broad Street in Soho around which a cluster of cholera cases had developed.

But he was widely ignored, and chlorine was being spread in the streets for the wrong reason: to combat the supposedly dangerous smell, not to kill water-borne germs. In the 'great stink' of 1858, when parliamentarians were so disgusted by the smell from the River Thames that they at last authorized the construction of modern sewers to carry the sewage out to sea, chloride of lime was applied to window blinds in Parliament to mask the smell.

So the source of the invention of chlorination, like that of vaccination, is enigmatic and confused. Only in retrospect can it be seen as a disruptive and successful innovation that saved millions of lives. It evolved rather slowly, probably from serendipitous beginnings in largely mistaken ideas.

How Pearl and Grace never put a foot wrong

In the 1920s the most lethal disease to affect American children was whooping cough, or pertussis. It killed about 6,000 children a year, more than each of diphtheria, measles and scarlet fever. There were vaccines for whooping cough available in some places but they were almost useless. The only preventive was quarantine and even this worked poorly since nobody knew how long it was necessary to isolate the victims for. It was this problem that drew a pair of ordinary but extraordinary women, both of whom had begun their careers as teachers, into research on the disease.

Pearl Kendrick, from New York State, had studied bacteriology in 1917 at Columbia University, while working as a teacher. By 1932 she was at the Michigan state public health laboratory in Grand Rapids, busily analysing the safety of water and milk. That year she recruited Grace Eldering, who was originally from Montana and had likewise turned from teaching to bacteriology, to join her team. At the time an outbreak of virulent whooping cough was sweeping through the city, and Kendrick asked her boss if she could work on it in her spare time. She and Eldering set out to develop a reliable test for who was infectious. This was a 'cough plate' of the medium in which the pertussis bacterium would grow, and on to which patients would cough. If the bacteria grew, then the patient was infectious.

Laboriously manufacturing their own medium for the cough plates and going out to homes all over Grand Rapids, to collect samples, at the end of long days of paid work, Kendrick and Eldering had their eyes opened to the deprivation that had worsened the plight of the working class since the start of the Depression. By the light of kerosene lamps, they saw children struggling for breath in workless and hungry households. Quarantine sometimes meant destitution for a family where the breadwinner could not then go out to work. Soon they had established that most people were infectious for four weeks, which helped influence local and national policy on quarantine. But they wanted to go further and develop an effective vaccine.

And this they did, systematically and gradually, by using standard vaccine development techniques, over the next four years: nothing new or really clever, just careful experiments. The end result was a killed version of several strains of the bacterium that, when injected into mice, guinea pigs, rabbits, and Kendrick's and Eldering's own arms, proved safe. Now to show that it protected people against whooping cough.

Here the two scientists proved to be adept at the social as well as the laboratory side of the work. They did not want to do what was typical at the time and use as a control group orphans who would be denied the vaccine to show that it worked, but they needed to match those who received the vaccine with similar people who did not receive it. With the help of local doctors and social workers they used the Kent County Welfare Relief Commission's statistics to identify a selection of people who matched those given the vaccine in age, sex and location but who had missed out on the vaccine for whatever reason. During 1934–5, they found that four of the 712 vaccinated children caught whooping cough, whereas forty-five of the 880 unvaccinated controls went down with the disease.

When Kendrick and Eldering announced these results at the annual meeting of the American Public Health

Association in October 1935, the audience was sceptical, suspecting that the trial was faulty in some way – as
many trials were in those days. A doubtful medical scientist named Wade Hampton Frost came twice from Johns
Hopkins to examine the methods employed but eventually admitted that he could find nothing wrong with the
two women's work. At the same time, Kendrick wrote to
Eleanor Roosevelt, inviting her to visit the laboratory, and
to her amazement received an acceptance. The First Lady
spent thirteen hours with the two scientists and went back
to Washington to persuade the administration to find ways
to fund the project so they could hire more people to help.
This enabled Kendrick and Eldering to do a second, larger
trial, using just three injections per child instead of four, to
which large numbers of families immediately applied. In
1938, when the second trial produced even stronger results,
Michigan began mass-producing the vaccine and by 1940
the rest of the country had followed suit, followed by the
rest of the world. Whooping cough incidence and mortality
fell rapidly and permanently to very low levels.

Kendrick and Eldering received very little recognition
for this work, turning down most requests from the media
even decades later, and little financial reward. They shared
their methods and formulae freely all over the world. They
did everything right: chose a vital problem, did the crucial experiments to solve it, worked with communities to
test it, gave it to the world and wasted no time or effort
defending their intellectual property. When not travelling
to spread the message and the vaccine, they lived together
in a house in Grand Rapids and threw generous parties
and picnics for their co-workers. Nobody had a bad word
to say of them. As one colleague said later: 'Dr. Kendrick

never became rich and, outside a relatively small circle of informed friends and colleagues, never became famous. All she did was save hundreds of thousands of lives at modest cost. Secure knowledge of that fact is the very best reward.'

Fleming's luck

Fifty years after Pasteur's summer holiday led to a fortuitous insight into the mode of action of vaccination, summer holidays would produce another piece of serendipity in the conquest of disease. Alexander Fleming left his laboratory in London to spend August 1928 in Suffolk. The weather that summer in London was changeable: cool in much of June, then suddenly rather hot in July, the temperature reaching a stifling 30°C on the 15th, before cooling dramatically in early August, then heating up again after 10 August. This is of relevance because it affected the growth of the *Staphylococcus aureus* bacteria that Fleming was growing in petri dishes as he prepared a chapter for a book on bacteria. Though he was an expert on the species, he wanted to check some of his facts. The cold spell of early August was just right for the growth of a mould, of the fungus *Penicillium*, a spore of which had somehow floated into the laboratory on the wind and landed on one of the petri dishes. The hot spell that followed allowed the bacterial culture to expand, leaving only a gap around the *Penicillium*, where the mould had killed the staphylococci. It created a striking pattern, as if the two species were allergic to each other. Had the weather been different this pattern might not have been possible, because penicillin is not effective against mature bacteria of this species.

Fleming, a diminutive and taciturn Scot, returned from holiday on 3 September and – as was his habit – began inspecting the cultures he had left behind in the petri dishes cluttered into an enamel tray, before discarding them. A former colleague, Merlin Pryce, put his head around the door and Fleming engaged him in conversation as he worked. 'That's funny,' he said when he picked up the plate with the pattern of exclusion between the fungus and the bacterium. Was the fungus producing a substance that killed bacteria? Fleming was immediately intrigued and saved both the plate and a sample of the fungus.

Yet it was to be more than twelve years before anybody turned this discovery into a practical cure for diseases. Part of the problem was the success of vaccination. Fleming's career had been largely under the influence of a great pioneer of bacteriology, Sir Almroth Wright, who was convinced that diseases would never be cured by medications, however effective, but by assisting the body to defend itself. Vaccination should be used to treat, as well as prevent disease.

Wright, the son of an Irish father and a Swedish mother, was a towering figure, outspoken, eloquent and irascible. Among colleagues he was known as 'the Praed Street Philosopher', 'the Paddington Plato' or more mischievously Sir Almost Right or Sir Always Wrong. 'Stimulate the phagocytes!' was Wright's battle cry, immortalized in Bernard Shaw's play *The Doctor's Dilemma,* in which Sir Colenso Ridgeon is a thinly disguised depiction of Wright. St Mary's Hospital, where Wright and Fleming worked, became the high temple of vaccine therapy. Wright's championing of typhoid vaccination for the Allied troops in the First World War probably saved hundreds of thousands of lives.

Influenced by Wright, Fleming's scepticism that a chemical could ever be found to cure infections was reinforced by his experiences researching the causes of infection in wounds during the First World War. He and Wright were stationed in a casino in Boulogne, which they turned into a bacteriological laboratory, the better to understand how to save lives. Here Fleming showed, using test tubes deformed to resemble jagged wounds, that antiseptics like carbolic acid were counterproductive, because they killed the body's own white blood cells without reaching the gangrene-causing bacteria deep in the crevices of the wounds. Instead, Fleming and Wright argued, wounds should be cleaned with saline solution. It was an important discovery, and one that doctors treating the wounded almost completely ignored, because it felt all wrong not to dress wounds with antiseptics.

Yet Fleming was not dogmatic about his devotion to Wright's ideas. Before the war he had adopted the medical scientist Paul Ehrlich's arsenic-based chemotherapy, Salvarsan, for syphilis and became renowned as a 'pox doctor'; so he knew that there were other ways to treat disease than by stimulating the phagocytes. In 1921 he discovered the bacteria-killing properties of a protein, lysozyme, found in his own nasal mucus and in tears, saliva and other fluids – and that was secreted by phagocytes. The body's own natural antiseptic, lysozyme hinted at the possibility of finding chemicals that could kill bacteria when injected into the body. But lysozyme itself proved disappointing against the most virulent species of bacteria responsible for disease.

Thus Fleming was at least partly prepared for the discovery of penicillin, or 'mould juice' as he initially called it. In a series of experiments he showed that it killed many kinds of virulent bacteria more effectively than most antiseptics, but did not kill the body's own defensive phagocytes. But early experiments with penicillin as a topical antiseptic applied to infected wounds were disappointing. Nobody yet realized that it worked best if injected into the body. Also, it was hard to produce in quantities or to store. Notoriously, in 1936, the pharmaceutical company Squibb concluded that 'in view of the slow development, lack of stability and slowness of bacterial action shown by penicillin, its production and marketing as a bactericide does not appear practicable.' So penicillin languished as a curiosity, undeveloped as a cure for disease, for more than a decade. Fleming was a denizen of the laboratory, not the clinic or the boardroom.

It is generally assumed that it was the outbreak of war that accelerated the development of antibiotics, but the evidence suggests that this may be wrong. On 6 September 1939, just three days after war broke out, two Oxford scientists applied for a grant to study penicillin, still thinking in terms of science, rather than application. They had been working on it for more than a year by now, and the outbreak of war actually made it harder for them to get the money. Both the Medical Research Council and the Rockefeller Foundation gave considerably less than requested, and the latter cited wartime uncertainty as the reason. So, if anything, the outbreak of war slowed down the development of penicillin at this stage.

The two scientists, Ernst Chain, a refugee biochemist from Germany, and Howard Florey, a pathologist from Australia, had come across Fleming's work and decided to take a closer look before war began. Despite wartime shortages of suitable materials, money and people, by May 1940 their colleague Norman Heatley had extracted penicillin and injected it into mice to show that it did not harm them. On Saturday, 25 May, Florey injected four mice with penicillin before infecting them, and four control mice, with a huge dose of streptococcus bacteria. That night the four untreated mice died; the treated mice survived.

It dawned on Florey, Chain and Heatley that a new treatment for wounded soldiers might be at hand. Over the next few months they turned their laboratory into a penicillin factory, and on 12 February 1941 Albert Alexander, a 43-year-old policeman who was dying of septicaemia following a scratch from a rose bush, became the first person to be treated with penicillin. He quickly rallied, but supplies of penicillin dried up before he was fully cured, and he relapsed and sadly died. But the miraculous effect of the drug had been observed. Then in August 1942 Fleming (whose interest had now been reawakened) used penicillin to cure one Harry Lambert of meningitis, in a case that caught the attention of the press. From then on Fleming was a hero, more so than the publicity-shy Florey.

In July 1941, with the war stretching British industry to breaking point, Florey and Heatley flew to America to kick-start the production of penicillin there. Higher-yielding varieties of mould were quickly discovered and better techniques for culturing them, but chemical companies were initially reluctant to invest in such an uncertain project, while anti-trust (i.e. anti-monopoly) rules made it hard for the firms to learn techniques from each other. Britain was later somewhat resentful of just how much intellectual property relating to penicillin was then claimed by American industry.

Wartime shortages, security concerns and – in Britain – V1 flying bombs continued to hamper the project, so it is by no means clear that penicillin would have been developed more slowly in peacetime. This is not to deny the drug's value to wounded soldiers, many of whose lives were saved. Even more remarkable was penicillin's ability to cure gonorrhoea, an enemy that was causing more casualties than the Germans in the North African and Sicilian campaigns. By D-Day enough penicillin was available to ensure that the death rate from wounds was far lower than expected.

News of penicillin's properties had reached Germany even before the war, and Hitler's doctor used it to treat the Führer after the assassination attempt of June 1944, but no serious attempt was made to scale up production there or in France. This too would probably have been different in a peaceful 1940s.

The story of penicillin reinforces the lesson that even when a scientific discovery is made, by serendipitous good fortune, it takes a lot of practical work to turn it into a useful innovation.

The pursuit of polio

By the 1950s the most high-profile disease in the United States was polio. The story of polio's vaccine is not quite as neat and harmless as that of smallpox. Some of the worries of the early opponents of Lady Mary Wortley Montagu did indeed come to pass, albeit much later. Vaccines did cause deaths that they should not have done. And it was another stubborn, unfashionable woman who blew the whistle.

Her name was Bernice Eddy. Born in rural West Virginia in 1903, the daughter of a doctor, she could not afford medical school, so she went into laboratory research, receiving a PhD in bacteriology in 1927 from the University of Cincinnati. By 1952 she was working on the polio virus in the Division of Biological Standards, a branch of the US government. She was involved in testing the new Salk vaccine for safety and efficacy.

Polio became a worsening epidemic especially in the United States during the twentieth century. Ironically, it was mainly improved public health that caused this, by raising the age at which most people caught the virus, resulting in more virulent infections and frequent paralysis. When everybody encountered sewage in their drinking or swimming water, the population was immunized early, before the virus caused paralysis. With chlorine cleaning up the water supply, people encountered the virus later and more virulently. By the 1950s the polio epidemic in the United States was worsening every year: 10,000 cases in 1940, 20,000 in 1945, 58,000 in 1952. Enormous public interest channelled generous donations into treatment, and the search for a vaccine. Huge fame and great wealth awaited the team that reached the prize, so some corners were cut.

One breakthrough came when Jonas Salk in Pittsburgh used the new technique of tissue culture to grow vast quantities of polio virus in the minced kidneys of monkeys. By 1953 he was killing fifty monkeys a week for their kidneys, growing the viruses in flasks of kidney tissue culture and inactivating them with thirteen days of exposure to formaldehyde. The vaccine thus produced was tested on 161 children and found to cause no harm, no polio and raised antibodies against the polio virus. Overcoming objections from Salk's rivals, especially Albert Sabin, and ignoring the results of Bernice Eddy, who found that the vaccine could still sometimes cause polio in monkeys, the Salk vaccine was rushed into nationwide trials in 1955 with a great fanfare of publicity. Disaster followed when one of the manufacturers, Cutter Laboratories, infected thousands and paralysed more than 200 people with polio through inadequately inactivated virus. The vaccine was quickly withdrawn and the programme rethought.

Meanwhile Dr Eddy had another concern. With Sarah Stewart she had done a groundbreaking experiment to show that cancer could be transmitted from a tumour in a mouse to a hamster, rabbit or guinea pig, by a virus, the SE polyoma virus (the 'S' standing for Stewart and the 'E' standing for Eddy) – a momentous biomedical discovery. She knew that the monkey kidney tissue culture used to grow the Salk vaccine sometimes itself sickened with viral infections, because of monkey viruses, and she worried that these contaminating viruses might be included with the vaccine, and might cause cancer in people. In June 1959 in her own time she did experiments to show that monkey kidney cultures could indeed cause cancers in hamsters, at the site of inoculation. She was rebuked by her boss, Joe Smadel, for doing the work, because of the way it cast another doubt on the safety of polio vaccination, and when she persisted in reporting it to a scientific meeting in October 1960, she was effectively sacked from polio work and forbidden to speak about her experiments. Smadel thundered: 'You have apparently stirred up a hornet's nest, and there are some who are sufficiently credulous to believe that the use of monkey kidney tissue cultures in man may induce cancer in them.' Indeed.

The contaminating virus was eventually isolated, christened SV40 and studied in detail by others. We now know that almost every single person vaccinated for polio in America between 1954 and 1963 was probably exposed to monkey viruses, of which SV40 – the fortieth to be described – was just one. That is about 100 million people. In the years that followed, the health establishment was quick to reassure the world that the risk was small, but they had little reason to be so complacent at the time. Sure enough there has been no epidemic of unusual cancer incidence among those who received contaminated vaccines, but SV40 DNA has been detected in human cancers, especially mesotheliomas and brain tumours, where it may have acted as a co-factor alongside other causes. Saying this remains unpopular to this day.

Polio eradication was targeted in 1988. Using a combination of inactivated vaccines to prevent paralysis and oral (live) polio vaccines to create full immunity, volunteers fanned out across the world to find and bring protection to adults and children everywhere. They continued throughout civil wars, crossing front lines, and even winning ceasefires to do their work, in wars in South America and central Africa. Over the next thirty years they probably prevented 16 million cases of paralysis and 1.6 million deaths. Today they have been more than 99.99 per cent successful. The last case of polio in Africa was in 2016. Only Afghanistan and Pakistan still report a very few cases: thirty-three in 2018. There, too, it will surely soon be history.

Mud huts and malaria

By the 1980s, with smallpox eradicated, and polio, typhoid and cholera in retreat, one stubborn disease remained the biggest killer, capable of ending hundreds of thousands of lives a year. And it was getting worse: malaria.

On 20 June 1983, in the hot and dusty settlement of Soumousso, in Burkina Faso, West Africa, a group of French and Vietnamese scientists began an experiment together with African colleagues. They had bought in the local market some tulle and percale cotton cloth with which to make thirty-six mosquito nets. Some were large group nets to cover more than one bed, some individual nets to cover a single bed. They now soaked half the nets in a 20 per cent solution of the insecticide permethrin and left the other nets untreated. They next did something rather odd: they tore lots of small holes in half the nets, both the treated and the untreated ones. They now had nine nets untreated and unholed, nine that were treated and unholed, nine that were treated and holed and nine that were untreated and holed. They then laid the thirty-six nets flat in the sun for ninety minutes to dry before installing them in twenty-four huts. These huts were built with the traditional mud walls and thatched roofs, but they were not meant to be used as homes. This being a research station, they were specially equipped with mosquito traps, some designed to catch the mosquitos inside the huts, and some to catch them leaving the huts.

On 27 June, volunteers began to sleep in the huts, occupying them from 8 p.m. to 6 a.m. every night for five months, one person to an individual net, three to a group net. Six days a week, three times a day, every mosquito that entered or tried to leave the huts was collected, dead or alive: at 5 a.m., 8 a.m. and 10 a.m. The live mosquitoes were kept under observation for twenty-four hours to see how many could be added to the roll call of the dead. After twenty-one weeks, 4,682 female mosquitoes had been collected, mostly of two species: *Anopheles gambiae* and *Anopheles funestus*, both malaria vectors.

The idea for this experiment had occurred to two of the French scientists, Frédéric Darriet and Pierre Carnevale, after noting the use of DDT-treated bed nets by the American military in the Second World War and by Chinese forces later. *Why include nets with holes in them?* I asked Darriet recently. Because mosquito nets rarely remain intact for long in Africa, so it is realistic to study whether a torn net is as useless as nothing or as useful as an intact net. In the case of untreated nets, a torn cloth is pretty useless, as many a restless sleeper will have experienced. But what if there is insecticide on the net to kill or repel the insects?

The Burkina Faso team's results were truly astonishing, even to Darriet and Carnevale. They found that the presence of a permethrin-treated net, whether intact or torn, repelled mosquitoes. It reduced the number of mosquitoes entering the huts by about 70 per cent and increased the rate at which the insects left the hut from 25 per cent to 97 per cent. And it reduced the 'engorgement rate' – whether the mosquitoes took a blood meal – by 20 per cent for *An. gambiae* and 10 per cent for *An. funestus*. Whereas hardly any of the mosquitoes in the control huts died, 17 per cent of those in the huts with treated nets died. After five months, the nets were still highly effective as insect repellents and killers. Today the treatment of nets lasts even longer.

This beautifully simple, carefully designed, experiment, known as 'Darriet et al. 1984', became famous in the small world of malaria and insect control, though it has never achieved the celebrity it deserves in the popular media. It proved to be a breakthrough in the control of malaria in Africa. The impregnated bednet is the magic bullet against the disease and its vectors. The idea took a while to catch on. Impregnated bednets first started to be used on a wide scale in 2003, and that very year malaria mortality stopped increasing and began to decline. According to a recent study published in *Nature*, insecticide-treated mosquito nets account for 70 per cent of the six million lives saved worldwide in recent years, twice as high a percentage as anti-malarial drugs and insecticide sprays put together. By 2010, 145 million nets were being delivered each year. Over a billon have been used to date. Globally, the death rate from malaria almost halved in the first seventeen years of the current century.

Tobacco and harm reduction

The greatest killer of the modern world is no longer a germ, but a habit: smoking. It directly kills more than six million people every year prematurely, perhaps contributing indirectly to another million deaths. The innovation of smoking, brought from the Americas to the Old World in the 1500s, is one of humankind's biggest mistakes.

Given that this is a voluntary habit, and that human beings are rational at least some of the time, it ought to be relatively easy to exterminate this killer. Just tell people it is bad for them and they will stop. Addiction being what it is, this has proved harder than that. Smoking is the source of more premature mortality than almost any other cause. Knowing that it causes cancer and heart disease made surprisingly little dent in its global popularity. The proof that smoking kills having long been established beyond all reasonable doubt has done surprisingly little to stop the habit. Advertising bans, plain packaging, bans on smoking in public spaces, deterrent messages on cigarette packages, medical advice, education – all have had some effect, especially in Western countries. But still more than a billion people in the world are addicted to lighting little bonfires of plant material between their lips.

Enter innovation. The decline of smoking in Britain has accelerated sharply in recent years, largely because of the spread of an alternative way of getting nicotine hits (which are not known to be harmful in themselves), using high technology instead of smoke: the electronic cigarette. More people vape in Britain than in any other European country. About 3.6 million Britons vape, compared with 5.9 million who smoke. The habit is even endorsed by public agencies, the government, charities and academic colleges, not because it is wholly safe, but because it is much safer than smoking. This is in sharp contrast to the United States, where vaping is officially discouraged, or Australia, where it is still – as of this writing – officially illegal.

Who was vaping's innovator? The original inventor was a man named Hon Lik who devised the first modern electronic cigarette in order to stop himself smoking. Around the turn of the twenty-first century, he was working as a chemist at the Liaoning Provincial Institute of Traditional Chinese Medicine and smoking two packs of cigarettes a day. He wanted to quit but tried and failed several times.

He tried a nicotine patch but found it a poor substitute for the hit he got from a cigarette.

One day in the laboratory at work he acquired some liquid nicotine and began experimenting with ways of vaporizing it. The first commercial electronic cigarette had been marketed in the 1980s, without success, and prototypes date back to the 1960s, with patents on the use of nicotine vapour even in the 1930s. Now with the miniaturization of electronics, however, Mr Hon had better luck. His first machine was big and cumbersome, but by 2003 he had filed a patent on a smaller device using a more practical mechanism. Further miniaturization followed and he submitted the product for testing at the Pharmaceutical Authority in Liaoning and by the Chinese military's medical institute. It went on sale in 2006. But, remember, the inventor is not necessarily the innovator. Vaping has not succeeded in catching on as much in China as in Britain. Why?

In 2010 Rory Sutherland, an advertising executive stopped by an office in Admiralty Arch in central London to see an old friend, David Halpern, who had just begun working as head of David Cameron's new Behavioural Insights Team, otherwise known as the 'nudge unit'. During the course of the conversation, Sutherland pulled out an electronic cigarette he had bought online and inhaled.

By then electronic cigarettes had been banned in Australia, Brazil and Saudi Arabia among other countries, at the urging of either the tobacco farmers or public-health pressure groups worried that this was in effect a new form of smoking. It was surely only a matter of time before Britain also outlawed the technology.

Halpern had not seen an electronic cigarette. He asked Sutherland to explain and was intrigued by the thought that the risks of vaping might be the lesser of two evils – like vaccinating to prevent smallpox or chlorinating to prevent typhoid. Or like distributing clean needles to heroin addicts to prevent HIV infection, a controversial policy adopted by Britain in the 1980s which had proved remarkably effective in keeping the HIV infection rate among drug addicts far lower than in other countries. 'We looked hard at the evidence and made a call,' wrote Halpern later. 'We minuted the PM and urged that the UK should move against banning e-cigs. Indeed, we went further. We argued we should deliberately seek to make e-cigs widely available, and to use regulation not to ban them but to improve their quality and reliability.'

That is why this innovation caught on more in Britain than elsewhere, despite furious opposition from much of the medical profession, the media, the World Health Organization and the European Commission. Strong evidence from well-controlled studies now exists that vaping's risks, though not zero, are far lower than smoking: it contains fewer dangerous chemicals and it causes fewer clinical symptoms. One 2016 study found that after just five days of vaping, the toxicants in the blood of smokers had dropped to the same levels as those of people who quit altogether. A 2018 study of 209 smokers who switched to e-cigarettes and were followed for two years found no evidence of any safety concerns or serious health complications.

But vaping ran up against the same sort of entrenched opposition from vested interests as greeted Lady Mary Wortley Montagu's inoculation. Tobacco interests got it banned in many countries; pharmaceutical companies lobbied to have it restricted in others, the better to protect their prescribed gums and patches; public-health lobbies argued against it, the better to protect their stop-smoking practices. In 2014, at the height of an ebola epidemic, which ought to have been a priority, the director-general of the World Health Organization, Margaret Chan, made it clear that she considered opposing vaping a high priority. The European Commission also tried to kill the industry in 2013 by demanding e-cigarettes be regulated as medicinal products.

That proposal was dropped, but Europe's Tobacco Products Directive, which came into force in 2017, brought a ban on high-strength e-liquids and on the advertising of e-cigarettes. This compromise partly helped the industry by introducing standards and subjecting products to strict product-safety regulations, including toxicological testing of the ingredients, as well as rules to ensure tamper-proof and leak-proof packaging. In the United States, by contrast, there was little regulation, but there were lots of attempts to ban vaping products, and sure enough people soon began to die, almost all of them as a result of buying black-market vaping products containing not nicotine, but THC oil, an ingredient of cannabis, contaminated with a thickening agent called vitamin E acetate. In effect, in an echo of the Prohibition era, while the British government encourages vaping but strictly regulates the products, the American government discourages it then does little to ensure its safety.

3

Transport

Failure is only the opportunity to begin
again more intelligently.

HENRY FORD

The locomotive and its line

For all of human history until the 1820s, nobody went
faster than the speed of a galloping horse. Then within a
generation it became routine to travel at speeds three times
that fast, and for hours at a time. Has there ever been an
innovation as tangible and dramatic as this? I have lived, by
contrast, in an era when speed of transport has not changed
much at all.

The man who did most to make the breakthrough in
speed was not the originator of the idea but the practical improver, and like Thomas Newcomen a craftsman
of humble origin. The year is 1810 and a new coal mine
has been sunk at Killingworth in Northumberland, with
a brand-new Newcomen engine installed to pump out the
water. But it does not work, and for a whole year the pit

remains drowned, despite the best efforts of engine-men from all around. In a tale very reminiscent of James Watt's, the humble brakesman in charge of the winding gear at a neighbouring pit, 29-year-old George Stephenson, who has a reputation for being able to mend clocks and shoes, offers to help. His only condition is that he will pick his own workmen to help him. Four days later, having dismantled the engine, reshaped the injection cap and shortened the cylinder, he has the engine working well, and the pit is soon dry. Stephenson gets the job of engineman and is soon much called upon as an engine doctor all over the district.

Stephenson's father was a 'fireman' at Wylam colliery, his job being to shovel coal into the furnace to fuel a steam engine. Young George quickly rose by seventeen years of age to being the 'plugman' in charge of such an engine at Newburn and then the 'brakesman' in charge of the steam-driven winding gear at Willington Quay and later Killingworth. He had by now been hit by a series of misfortunes: his wife died, leaving him with a young son. His father was blinded by a steam engine accident. He was drawn for military service and had to buy his way out by paying for a substitute to serve in his place, using up the last of his savings. But with his mechanical reputation growing he was soon in great demand. And the time was ripe for locomotion.

The idea of a steam engine pulling wagons along a wagonway was not new. Stationary engines had been hauling coal-wagons up hills for some years now using cables, and Richard Trevithick's first locomotive steam engine had hauled a train along a track in Merthyr Tydfil in 1804. Trevithick realized that high-pressure steam could now be handled by modern metalworking, giving much more power, making an engine portable and doing away with the need for a condenser. But Trevithick could not make money, lost interest, travelled abroad and died penniless; the experiment seemed to be over. His imitators likewise gradually gave up. Steam locomotives were unreliable, dangerous, hideously expensive, damaging to wooden or iron-plate railways and unable to haul heavy loads or go up hills without their wheels slipping. Better to stick to horses, said everybody sensible.

What changed this was war. The Napoleonic conflict created an insatiable demand for horses and for hay to feed them, driving up the price of both. In the coal-mining districts, getting the coal to water by horse-drawn cart to be loaded on to ships was the limiting factor – a journey of more than eight miles would make a pit unprofitable. So pit owners started experiments again, and all over the northeast there were clanking machines with boilers aboard trying to get up speed. Even so, almost nobody imagined that the railway would prove useful outside the world of the collieries. That it could challenge the canals and the stagecoaches to carry people and cargo long distances just did not occur to anybody: an illustration of a great truth about innovation, that people underestimate its long-term impact.

In 1812 an ingenious engineer named Matthew Murray in Leeds built an engine for John Blenkinsop with two cylinders rather than one, called *Salamanca* after the Spanish battle in which Arthur, Lord Wellington, defeated a Napoleonic army. He then shipped another of the same design called, by spelling mistake, *Willington*, to the northeast. It progressed forward using a cog, rack and pinion system, but William Hedley's rival *Puffing Billy* at Wylam

(1813) did away with this, disposing finally of the persistent myth that a smooth wheel could not grip a smooth rail.
Counterintuitively, with enough weight a locomotive can
pull a heavy load on the smoothest of rails at least on a
shallow incline. But Hedley and others soon encountered
a new problem: the iron plates of the wagonway could
not handle the weight of a locomotive and kept smashing.
Clearly some innovation was needed under the wheel as
well as on top of it.

This was the moment for Stephenson, who more than
anybody else saw the need for innovation in both engine
and rail. The next year, 1814, he built a two-cylinder locomotive at Killingworth, named *Blücher* after the victorious
Prussian general (the influence of the Napoleonic War
in this story continues). He largely copied the design of
Murray's *Willington*. When it was working, *Blücher* proved
capable of hauling fourteen wagons carrying 2 tons of coal
each at 3 miles an hour, doing the work of fourteen horses.
It was not reliable enough yet to compete with horse-drawn
vehicles, let alone canals, except at coal mines, where fuel
was cheap, but it was a start. And Stephenson was already
tinkering with the design.

As for rails, Stephenson soon took out a patent with
William Losh on a new design of cast-iron rail, better able
to withstand the weight of the locomotive. But he then
changed tack. A friend named Michael Longridge had
recently taken over running an ironworks at Bedlington
on the River Blyth, not far from Killingworth, and, using
the new puddling process for producing malleable (low-
carbon) iron, had the idea of rolling out wrought-iron rails
through a mould. Longridge's engineer, John Birkenshaw,
came up with a design of rail that was wedge-shaped in

cross section, with a broad top and a narrow base. This saved metal while allowing good contact with the wheel of a locomotive. When it came to the building of the Stockton to Darlington railway in 1822 (I am writing these words in Darlington railway station!), Stephenson abandoned cast iron, to the fury of Losh, and went with Birkenshaw's wrought-iron rails.

George Stephenson and his son, Robert, now did something astonishingly bold. They surveyed and built a twenty-five-mile (eventually forty-mile) wrought-iron railway equipped with locomotives to haul coal to Stockton from Darlington. Serendipity played a part in this triumph. The wealthy Quaker wool merchant and philanthropist Edward Pease had proposed a horse-drawn railway, rather than a canal, to get coal, wool and linen from Darlington to the tide at Stockton-on-Tees. But building even a horse-powered railway, like building a canal, required an immense expenditure on lawyers and agents to acquire land and get an Act through Parliament. Pease and his fellow Darlington Quakers encountered fierce opposition from members of the House of Lords and it was only Edward Pease's single-minded determination and hard slog in bending the ears of politicians in London that finally got a bill passed in April 1821. This was just for a horse-drawn railway.

The very day the bill passed, 19 April 1821, Pease met George Stephenson who had walked to see him from Stockton, having apparently heard that he was planning a railway. Stephenson offered to survey the route, and then persuaded Pease to use locomotives as well as horses. This led to a fresh round of fury among the landowners, terrified by 'ridiculous' rumours (said the promoters) that these 'infernal machines' (said the opponents) might go at 10 or 12 miles an hour!

Robert Stephenson organized the construction of improved locomotives to run on the Stockton and Darlington railway. At the grand opening on 27 September 1825, the first of these, *Locomotion*, designed mostly by Timothy Hackworth, hauled a train that consisted of twelve wagons of coal, one of flour, and twenty-one of people. By the time it reached Stockton, more than 600 people were aboard. Later *Locomotion* showed it could travel at up to 24 miles an hour. Heat was doing work in transporting people for the first time.

True, the Darlington and Stockton railway relied heavily on horses for the next few years, locomotives being occasional, unreliable and dangerous interlopers. But the Stephensons did not stop there. Their most famous locomotive design, *Rocket*, entered the Rainhill trials in 1829, a contest to choose engines for the Liverpool to Manchester railway, a line George Stephenson was building. To qualify, an engine had to weigh no more than 4.5 tons, have only four wheels, be well sprung, and haul a small train back and forth for thirty-five miles without stopping for any length of time – and then repeat the feat.

Rocket was designed by Robert, but incorporated many ingenious improvements, mostly invented by Henry Booth, a new collaborator. These included multiple fire tubes in the boiler to increase the rate of steam generation, angled cylinders, pistons connected directly to just two driving wheels and a blast-pipe for exhausting steam vertically into the chimney, increasing the draught through the furnace. In short, it was the product of incremental tinkering and trial and error by several people, not of brilliant leaps of imagination by a genius. At Rainhill, *Rocket* had nine competitors, five of which failed to start. Of the others, the horse-powered *Cycloped* collapsed, *Perseverance* failed, *Sans Pareil* cracked a cylinder, and *Novelty*, the crowd's favourite, set off at a furious lick but then kept bursting pipes. Playing tortoise to *Novelty*'s hare, *Rocket* steamed serenely on, hauling 13 tons and reaching 30 miles an hour. It set the basic design for locomotives for decades to come. It also caused the first railway fatality a year later, killing the politician William Huskisson at the grand opening of the railway, as he stepped off another train to speak to his political rival, the prime minister, the Duke of Wellington.

After the Liverpool–Manchester line was opened and proved a roaring success, nothing much happened for a few years: a scattered sprinkling of short railways appeared and techniques were slowly honed. Then, driven by low interest rates on government bonds and a liberated stock market, in 1840 there began an extraordinary boom in railway projects, funded by a frenzy of share-buying on deposit by anybody with savings. New rail lines appeared all over the nation, linking cities, then towns, then villages. Travel by rail became routine, fast and even a little bit reliable, though far from safe by today's standards. The grass grew between the cobblestones of roads as the stagecoaches vanished. The railway boom was a competitive bubble, profitable to some, ruinous to many, and riddled with hype and fraud, but enormously valuable to its users as it left Britain connected as never before, enabling trade to flourish.

The rest of the world soon followed suit. The first railway in America began operating in 1828, in France in 1830, in Belgium and Germany in 1835, in Canada in 1836, in India, Cuba and Russia in 1837, in the Netherlands in 1839. By 1840 America already had 2,700 miles of railway, and 8,750 by 1850.

Turning the screw

Putting steam engines on ships happened around the same time, but it was not until the second half of the 1800s, and the invention of the screw propeller to replace the paddle wheel, that ocean-going steam could challenge sail for price as well as speed. Sailing technology peaked in the late 1860s with the launch of the *Cutty Sark* and other fast clippers.

The story of the screw propeller shows all the usual elements of an innovation: a long prehistory, simultaneous breakthroughs by two rivals, then incremental evolution over many years. The idea had actually been around since the 1600s and it kept popping up in the eighteenth century, but by the 1830s paddle steamers were everywhere instead. Patent after patent appeared for screw designs – one historian traced 470 names associated with the idea, including an especially prescient patent in 1838 by a woman, Henrietta Vansittart, the mistress of the author Edward Bulwer Lytton – but practical trials were largely missing.

Then in 1835 a 27-year-old farmer in Hendon, on the outskirts of London, by the name of Francis Smith, built a model boat with a screw actuated by a spring and tried it on a pond. The next year he built a better one and took out a patent, on 'propelling vessels by means of a screw revolving beneath the water'.

By one of those remarkable coincidences, just six weeks later, also in London, a Swedish engineer by the name of

John Ericsson, who did not know Smith, also took out a patent for a similar device. Smith was already building a full-scale boat, of 10 tonnes, with a 6-horsepower engine, with the help of an engineer named Thomas Pilgrim. The boat was launched into the Paddington canal in November 1836 and immediately suffered a lucky accident. The screw that Smith had built was like a wooden corkscrew wrapped around a wooden shaft, with two complete turns of the screw along its length. A collision knocked one turn of the screw off, after which the boat went much faster, an unexpected discovery related to turbulence and drag. The next year Smith redesigned the propeller in metal with a single turn of screw and the boat went out to sea and round the Kent coast and back, proving its worth in rough weather. Ericsson's version had two drums, rather than a narrow shaft, with screws revolving in opposite directions, an adaptation that was largely unnecessary until the development of the torpedo.

Like most inventors, Smith struggled to be taken seriously. The Admiralty asked for a demonstration with a larger vessel, capable of at least 5 knots, before it would consider trying the technology. Smith formed a company, built a 237-ton ship called the *Archimedes*, fitted it with an 80-horsepower steam engine, and in October 1839 successfully took on the *Widgeon* at Dover and the *Vulcan* at Portsmouth, two of the Navy's fastest paddle steamers. Still the admirals demurred, while the *Archimedes* shuttled around Europe showing off. Eventually in 1841 the Admiralty commissioned a screw ship, *Rattler*, launched in 1843 and in service the following year. In 1845 *Rattler* was pitted against a paddle steamer of similar weight and horsepower, *Alecto*, in a tug of war, the two ships being attached by a line astern. *Alecto* was humiliatingly dragged backwards at 2 knots.

Meanwhile, in America, Ericsson had built a series of ships, including the *Princeton* for the US Navy. France had launched the screw-driven *Napoléon*. The world's navies switched to screws almost overnight. Innovation continued, though, and the design of the screw evolved radically as the years went by, and as the understanding of turbulence and drag improved. The blade shape eventually became narrow near the shaft, wide further out, then tapering to a rounded end.

Internal combustion's comeback

The story of the internal-combustion engine displays the usual features of an innovation: a long and deep pre-history characterized by failure; a shorter period marked by an improvement in affordability characterized by simultaneous patenting and rivalries; and a subsequent story of evolutionary improvement by trial and error. In 1807 a Franco-Swiss artillery officer not only patented but built a machine that could use explosions to generate movement. Isaac de Rivaz built a wheeled 'charette' on which was mounted a vertical cylinder in which hydrogen and oxygen were mixed and exploded by spark ignition: the weight of the descending cylinder drove the charette forward through a system of pulleys, before the explosions sent the piston back up again. It worked, just, as did a much bigger version built seven years later, but could not hope to compete with the steam locomotive.

In 1860, the year after the first oil well was sunk in Pennsylvania, Jean Joseph Étienne Lenoir patented a design for an internal-combustion engine running on petroleum and by 1863 he had built one that trundled very slowly for nine kilometres in three hours outside Paris. Called the hippomobile, it was a cart mounted on a tricycle. Its extremely wasteful inefficiency derived mainly from the fact that there was no compression of the air in the cylinder.

Two failures then. External combustion, to make steam, remained dominant for transport and would surely soon conquer the roads as well the rails. By the 1880s firms were springing up all over America and Europe to manufacture and sell steam cars, and as the new century dawned the main threat to the dominance of steam in the motor market would seem to be from new-fangled electric cars. The Stanley Steamer, marketed first in 1896 was the best-seller and set a world speed record of 127 mph ten years later. Yet within a few years, the underdog that was the internal-combustion engine had defied the experts and conquered all. Steam cars and electric cars were consigned to history.

The central invention behind internal combustion was the Otto cycle of compression and ignition, a four-stroke dance: fuel and air enter the cylinder (1), the piston compresses the mixture (2); ignition drives the power stroke (3) and the gases are exhausted by the piston (4). Nikolaus Otto, a grocery salesman, came up with this design in 1876 after sixteen years of trying to improve on Lenoir's engine. He had enough success along the way to make and sell stationary engines and to expand his firm, which became Deutz – still a leading engine maker.

Although Otto sold many engines, he was not interested in developing a car, so two of his employees, Gottlieb Daimler and Wilhelm Maybach, left and started making gasoline (petrol) engines for cars. Many others, in France, Britain and elsewhere contributed inventions during the 1880s, but it was Karl Benz who got the first complete car into series production in 1886. Benz, a talented engineer living in southern Germany and working at the back of a bicycle shop, built a three-wheeler to a design that owed more to bicycles than carriages. According to family legend, in 1888 his wife, Bertha, took the car without telling Karl, put her two sons in it and drove very slowly all the way from Mannheim to Pforzheim refuelling with gasoline bought from pharmacies along the way: a journey of nearly 100 kilometres. By 1894 more than 100 Benz *Motorwagen* had been sold.

Maybach and Daimler, meanwhile, were independently perfecting a four-stroke engine that ran at faster speeds than Benz's and gave much more power. In France, Émile Levassor acquired a licence to make Daimler engines and quickly began to make innovations in car design that were in turn copied by Daimler: the front-mounted engine and the water radiator among them.

In 1900 Maybach and Paul Daimler, son of Gottlieb (who had died that year), launched the car that set the standard of design followed by the industry ever after. The prototype was built specifically for Emil Jellinek, a wealthy Hungarian car racer living in Nice. Called the Mercedes 35hp, after the nickname of Jellinek's daughter, it no longer looked like the result of an encounter between a horse carriage and a steam engine round the back of a bicycle shed. Maybach made it wider, lower and with a low centre of gravity, to stop it overturning. It had an aluminium engine, mounted on a steel chassis over the front axle for the first time, and a patented water-cooled honeycomb radiator and gate gearbox. The car performed so well for Jellinek at Nice in 1901 that everybody wanted one, and the production plant in Stuttgart went flat out over the next few years.

Yet the Jellinek episode is a reminder that in the early years of the car industry, as in the early years of computers, mobile phones and many other innovations, the inventors thought they were developing a luxury good for the upper-middle classes. It took a farmer's son from Detroit to turn the car from a luxury invention into an everyman innovation: an affordable utility for ordinary people. Henry Ford revolutionized the industry after 1908, consigned steam and electric cars to history and brought cars within the reach of the masses, changing the behaviour of human beings in so many far-reaching ways that the automobile, not the aeroplane, is the twentieth century's representative technology, as the steam engine was the nineteenth's.

At the start, the eccentric and single-minded Ford would have seemed an irrelevant also-ran. He developed little that was technically new. He had twice set up car companies that failed, merely to try to emulate the expensive German and French designs: he abandoned the first and was sacked from the second. His third attempt, with an unremarkable design called the Model A, almost ran out of money, but sold just enough cars to keep going. But he had a relentless genius for cost control and he then began making a car that was simpler than most on the market, relatively cheap – and going to get cheaper with mass production. The Model T, the famous 'Tin Lizzie', was robust and reliable enough to appeal to Midwestern farmers needing to get to town. By 1909 he was selling them as fast as the factory could make them and was thinking big. With so few paved roads, the main competition was still the horse. As the Ford company argued in one of its advertisements: 'Old Dobbin, the family coach horse, weighs more than a Ford car. But – he has only one-twentieth the strength of a Ford car – cannot go as fast nor as far – costs more to maintain – almost as much to acquire.'

So who invented the motor car running on an internal-combustion engine? Like the steam engine and (as I will show later) the computer, there is no simple answer. Ford made it ubiquitous and cheap; Maybach gave it all its familiar features; Levassor provided crucial changes; Daimler got it running properly; Benz made it run on petrol; Otto devised the engine's cycle; Lenoir made the first crude version; and de Rivaz presaged its history. And yet even this complicated history leaves out many other names: James Atkinson, Edward Butler, Rudolf Diesel, Armand Peugeot and many more. Innovation is not an individual phenomenon, but a collective, incremental and messy network phenomenon.

The success of the internal-combustion engine is mainly a thermodynamic one. As Vaclav Smil has argued, the key metric is grams per watt (g/W): how much mass it takes to generate a certain amount of energy. Human beings and draught animals operate at about 1,000 g/W. Steam engines got that down to about 100 g/W. The Mercedes 35hp was more like 8.5 g/W and the Model T Ford just 5 g/W. And the cost just kept on falling. In 1913 somebody earning the average American wage would have had to work 2,625 hours to earn enough to buy a Model T. In 2013, on the average wage, he or she would have needed to work just 501 hours, or 18 per cent as long, to afford to buy a Ford Fiesta equipped with seat belts, air bags, side windows, rear-view mirror, heating, a speedometer and windscreen wipers – none of which were in a Model T.

The tragedy and triumph of diesel

Rudolf Diesel is an unusual innovation hero in several ways. He did not live to see the success of his device, apparently committing suicide by jumping from a ferry in the North Sea one night in 1913, while on the way to opening a Diesel factory in Britain, leaving large debts behind. He was motivated as much by social justice as ambition, believing (wrongly) that he was inventing something that would decentralize industry by being used in small machines, even sewing machines. 'My chief accomplishment is that I have solved the social question,' he said, after writing an unsuccessful book on how to organize worker-run factories. And unlike many inventors, Diesel started from first scientific principles. He became obsessed with the thermodynamics of the Carnot cycle, a theoretical idea by which an internal-combustion engine could reach 100 per cent efficiency, turning heat into work without changing temperature. In the 1890s he strove to get some way towards this goal by inventing an engine that used excess air and high compression, so that the fuel was ignited purely by compression, rather than a spark.

Neither of these ideas was novel, but Diesel's practical exploration of their possibilities eventually broke new ground. By 1897, thanks to the help of a more practical industrial engineer, Heinrich von Buz, he had a design of engine that worked at double the efficiency of the best gasoline engines then on the market, though it had largely abandoned many Carnot cycle features. At which point he and Buz thought they were home and dry. But getting to a reliable and affordable product proved almost impossibly difficult, mainly because of the challenge of making a machine that operated at high pressures. Diesel's critics said he had both claimed too much originality for his ideas and failed to make them workable. His disillusionment with life poured out in a letter written shortly before his death: 'The introduction [of an invention] is a time fraught with combating stupidity and jealousy, inertia and venom, furtive resistance and an open conflict of interests, an appalling time spent battling with people, a martyrdom to be overcome, even if the invention is a success.'

Yet, today, diesel engines run the world. Vast diesels – the biggest of which generate more than 100,000 horsepower – drive almost all the world's large cargo ships, making global trade possible, playing a larger role in globalization, argues Vaclav Smil, than political agreements over trade. Smaller diesels transport goods by road and rail; and virtually every farm tractor or bulldozer runs on diesel, making the modern economy unimaginable without them. In Europe in the early twenty-first century diesels even came to dominate the car market for a while after their efficiency appealed to politicians concerned about climate change, a decision that had to be reversed when the effect on urban air quality emerged.

The Wright stuff

Five years before the first Model T, in the month of December 1903, on the East Coast of the United States, after years of experiment, accident and disappointment, a human being was about to experience powered flight.

The American government had spent $50,000, through the War Department, to support the experiments of Samuel Langley, who was convinced that he could build a plane.

Another $20,000 had been contributed by Alexander Graham Bell, inventor of the telephone, and other friends of the aviation pioneer. Professor Langley, an astronomer, was a well-connected but rather haughty New Englander who was head of the Smithsonian Institution in Washington. He insisted on complete secrecy about the details of his device, sharing his ideas with nobody outside a small circle, but the demonstration had attracted a large crowd. Called the 'great aerodrome', his monstrous contraption, with a 48-foot wingspan, was to be launched from a track on the roof of a houseboat on the Potomac river, its gasoline-powered propeller driving it forward through the air while its two pairs of angled wings generated lift. Seven years before, in 1896, a model version with a steam-powered engine and no pilot had managed a promising 1,000 yards and ninety seconds of flight before crashing into the river. In August a repeat of that attempt had failed and in October the full-scale machine with a man aboard had simply dropped ignominiously straight into the water. This December test was probably Langley's last chance, but he was confident of success.

Not that Langley himself would be the pilot; he was too grand for that. That dubious privilege fell to Charles Manly, who at 4 p.m. climbed aboard the aerodrome, pessimistically wearing a cork-lined lifejacket. He powered up the engine and, after making some adjustments, blasted forward as the watching crowd held their breath. The machine curved straight up into the air, stalled, flipped backwards, began to disintegrate and crashed into the ice-flecked river not ten yards from the houseboat. Manly clambered from the wreckage, cursing volubly. Langley's reputation never recovered.

The fiasco led to an abrupt end to the government's support of powered flight after a decade of wasted money. Yet just nine days later, a few hundred miles to the south, on a windswept and sandy shoreline near a lonely fishing settlement called Kitty Hawk, with almost nobody watching, two brothers from Ohio would indeed achieve the first powered, controlled flight after spending a tiny fraction of Langley's budget. At 10.35 a.m. on 17 December 1903, with Orville Wright lying on the lower of two wings to control the steering gear and his elder brother, Wilbur, running alongside to steady the craft during take-off, the 'Flyer' lifted smoothly from a wooden track into a stiff headwind, its gasoline engine providing thrust and its biplane wings lift. Twelve seconds and forty yards later, it touched down on its skis. Just five people were watching. Later that day Wilbur flew the flyer for almost a minute and covered more than 800 feet.

Where Langley had done everything wrong – spending lots of money, depending on the government, consulting few other people, building a fully fledged device from scratch, rather than inching incrementally through each of the problems to be solved – the Wrights had done everything right. As experienced bicycle makers and diligent craftsmen, they had systematically worked step by step through the challenges necessary to solve the problem of powered flight. First they had written to and drawn upon the experience of others, especially the German designer of gliders, Otto Lillienthal (who died crashing one of his gliders in 1896), and an eccentric French-American in Chicago named Octave Chanute who had made a great study of the problems to be overcome and was himself a node in a huge network of exchanging ideas about flight. The Wrights sent Chanute 177 letters in all. The brothers also watched soaring birds obsessively. From all this research they had gleaned crucial ideas such as the curvature of an aerofoil wing to provide lift, the concept of a biplane and the notion of warping the wings to steer. Then in 1900 they had built a glider, taken it in parts to the windy Carolina barrier islands and tried it out, at first flying it tethered like a kite and then lying on it as it lifted into the wind while being run downhill. In 1901, despite a plague of mosquitoes and stormy weather, they camped at Kitty Hawk with two helpers and Chanute himself, having made adjustments to the design, only to find that it worked less well than the year before. The machine climbed quickly but stalled too easily. It turned out that Lillienthal's recommended ratio of the height to width of a curved wing – 1:12 – (which they had copied) was much too curved. With flatter wings, 1:20, the glider worked again.

At this point, back in Dayton, they began to experiment with models in a wind tunnel, making thousands of laborious measurements till they had a full understanding of lift and drag. As soon as the peak of the new bicycle-selling season was over in the summer, they returned to Kitty Hawk in 1902 with a third design of glider/kite, made more adjustments, especially to the rudder, and learned the hard way how to pilot a device through the air, crashing frequently till they mastered the art. Bit by bit they had put together everything except the motor.

Until now they had done nothing that could not at least in theory have been done by Leonardo da Vinci. A wooden frame with fabric covering constituted their invention. Sure, it had metal wires holding it in shape, and a sewing machine (invented in the Wrights' own state of Ohio not long before) was indispensable in making and repairing the wings. But this was just a sort of wooden hang-glider with huge wings and minimal weight, just capable of carrying a person. And it was pretty useless for any practical purpose, requiring strong wind for lift-off, but being easily blown into a crash. The reason nobody had invented such a thing before was partly because getting to the next stage, powered flight, had never been so tantalizingly close. The Wrights, surrounded by new-fangled automobiles, knew that it was the engine, the motor, that was going to make all the difference. Unlike other inventors they had left the motor to last, reasoning that it would be the least difficult thing to do because all it had to do was provide sufficient propulsion.

Here they had a stroke of luck. The man they had hired to run the bicycle shop while they were away, Charlie Taylor, was a very good mechanic. He could not find a lightweight engine on the market, so he designed and built one from scratch, using aluminium. It was a four-cylinder motor, and although it kept going wrong, eventually he had a version that proved reliable. Meanwhile Orville and Wilbur tinkered with different designs for a propeller, finding the mathematics fiendishly difficult and the example of a ship's propeller not especially helpful. By the autumn of 1903 all was ready. They moved to Kitty Hawk and late in the season managed at last to get into the air with a man lying behind an engine.

Most aviation pioneers, like Langley, were dilettanti gentlemen or scientists, rather than practical craftsmen. A distinguishing feature of the Wright brothers, who lived together with their preacher father, Milton, and their teacher sister, Katharine, was their dedication to hard work. Unmarried, uninterested in frivolity or anything remotely resembling sin, the siblings devoted their lives to working all the hours God gave, except on Sundays. They had each other as sounding boards, including Katharine, the only one with a university degree. In the photograph of the first flight, Wilbur, despite having spent many weeks in a makeshift camp and hangar on the North Carolina shore in freezing winds, is wearing a stiff collar with his black suit as if ready for church. As John Daniels, the Kitty Hawk resident who took the photograph, put it, they were the 'workingest boys I ever knew . . . It wasn't luck that made them fly; it was hard work and common sense.'

Even after the first flight, news of which was largely ignored by the world as highly implausible given the humble, non-graduate nature of the inventors, the Wright brothers continued to tinker with and tune their designs, till they were able to take off without a headwind using a catapult, turn slow circles in the air and stay up for minutes at a time. By 1905, at a field outside Dayton, Ohio, Wilbur had set a record of twenty-four miles of continuous flying. Yet even the local newspapers still did not catch on to what was happening under their noses, while the grandee commentators at *Scientific American*, even as late as 1906, saw fit to dismiss rumours about the claims of the brothers with patrician sarcasm in an article entitled 'The Wright aeroplane and its fabled performances':

> If such sensational and tremendously important experiments are being conducted in a not very remote part of the country, on a subject in which almost everybody feels the most profound interest, is it possible to believe that

the enterprising American reporter . . . would not have ascertained all about them . . . long ago?

Apparently so. Even when people did believe the Wright brothers, they doubted the value of what they had done. 'Our skepticism is only as to the utilitarian value of any present or possible achievement of the aeroplane. We do not believe it will ever be a commercial vehicle at all,' opined the *Engineering Magazine*.

The United States War Department replied to the Wright brothers' offer to demonstrate their flyer with a flat refusal, its mind made up by the Langley fiasco. When Wilbur travelled to France in 1907 and 1908 having signed a lucrative contract that would pay out if he could demonstrate powered flight and meet certain targets, he was still widely mocked as a bluffer. On the day set for his demonstration at a horse-racing track at Le Mans, on 8 August 1908, a small crowd gathered which included the grandee sceptic Ernest Archdeacon of the Aero Club de France, who continued to scorn the claims of the Wrights to all who would listen even while the crowd waited. Hours went by as Wright prepared the machine, and scepticism grew. The shock and excitement when he finally powered into the air at 6.30 p.m. was extreme. He turned to the left, flew back past the crowd, turned another circle and dropped smoothly back on to the grass after two minutes aloft at about thirty-five feet. 'The enthusiasm was indescribable,' said *Le Figaro*. 'C'est merveilleux!' cried Louis Blériot, who was there. 'Il n'est pas bluffeur!' shouted somebody, perhaps at M. Archdeacon.

Meanwhile at Fort Myer near Washington, Orville was also wowing the crowds with a duplicate flying machine.

On 9 September he twice stayed in the air for more than an hour, circling the field more than fifty times. From then on the two Wrights were huge celebrities, feted wherever they went. Their rivals scrambled to catch up; within a year twenty-two pilots went aloft at an air festival in Reims, watched by a crowd of 200,000, and Blériot had crossed the English Channel in a flimsy monoplane. Just ten years later, in June 1919, John Alcock and Arthur Brown crossed the Atlantic non-stop from Nova Scotia to Ireland in sixteen hours, through fog, snow and rain. The First World War had by then given rapid impetus to the development of designs and flying skills, though much of it would have happened anyway.

A long slog was still needed to turn the invention of powered flight into an innovation of use to society. Some of the Wrights' ideas were dropped: a forward elevator proved too unstable, and warping the whole wing to steer worked less well than having hinged flaps or ailerons. But their general discovery that to control an aircraft in a turn it was necessary to use the wings to achieve the roll and the rudder to control the yaw was crucial. The Wrights themselves soon became rich through prizes and contracts, yet were also embroiled in exhausting legal battles as they sought to defend their patents. Wilbur died of typhoid in 1912, at the age of forty-five. Katharine died in 1929, Orville in 1948.

Looking back, the Kitty Hawk moment of 1903 was bound to stand out, because there was only ever going to be one instant when a powered plane left the ground in controlled conditions, but in truth it was a step in a lengthy evolutionary path that began with strange, usually fatal attempts by eccentrics to leap into the air with big flapping wings. Likewise, the Wrights' design continued to evolve gradually into the airliners, supersonic jets, helicopters and drones of today. It is a continuum.

There is little doubt that somebody would have got planes into the air within the first decade of the twentieth century even without the Wrights. Motors made it inevitable that many people would then try, and trial and error was all that was really needed. Indeed, because so few people believed the Wrights at first (the *Paris Herald* called them 'Fliers or Liars' in 1906), the rival pioneers of powered flight, in France especially, such as Clément Ader, Alberto Santos-Dumont, Henri Farman and Louis Blériot, had begun quite independently hopping off the ground with propellers and wings more or less effectively and with more or less control.

To Orville Wright's fury, the Smithsonian tried to rewrite history in 1914, resurrecting Langley's aerodrome, secretly modifying it, flying it briefly, then removing the modifications before putting it on display along with the claim that Langley had therefore designed the first machine capable of powered flight. The Wrights' flyer was not installed in the Smithsonian museum until 1948, after Orville's death.

International rivalry and the jet engine

'The turbine is the most efficient prime mover known [so] it is possible that it will be developed for aircraft, especially if some means of driving [it] by petrol could be devised,' wrote young Frank Whittle in 1928 in a thesis on future aircraft design. He soon followed this, in 1930, with a patent on his own design for a jet. By that date the idea of the jet already had a fairly long history, including a patent taken out on a design for an axial-flow turbojet engine for planes in France by Maxime Guillaume in 1921, which Whittle did not know about. Bigger gas turbines were already in use running factories in France and Germany before this date, though they were far too inefficient to be adapted for flight.

But coming up with the idea of jet propulsion was a very long way from actually building a jet plane, as Whittle was about to find out. Finding materials for compressors and turbine blades that could withstand immensely high pressure and red-hot temperature while rotating at high speed was a tall order. As was the case with the steam engine in the 1700s, and is the case with nuclear fusion today, innovation in materials is vital to realizing an advance that can be conceived but not built. The engineer Alan Griffith had already been secretly wrestling with this problem at Britain's Royal Aircraft Establishment, since 1926. Griffith had published a key paper, 'An Aerodynamic Theory of Turbine Design', in that year, explaining the poor performance of all turbines: the blade shape was wrong and they were 'flying stalled'. Aerofoil shapes – of the kind used by the Wright brothers – proved much better. Griffith was now trying to come up with an axial-flow turbojet to drive a propeller in a two-stage engine, a forerunner of the turbo-prop.

When Whittle, a newly commissioned pilot officer, approached him, Griffith was welcoming but mildly discouraging, writing that 'the performance of both compressors and turbines will have to be greatly improved' before a jet would work. Whittle remembered this differently much later as a snub, yet the Royal Air Force generously sent Whittle to Cambridge to study engineering, and it was from there that he wrote to a friend in May 1935 saying that

'I have allowed the patent to lapse. Nobody would touch it on account of the enormous cost of the experimental work, and I don't think they were far wrong, though I still have every faith in the invention.'

Just six months later, in November 1935, Hans Joachim Pabst von Ohain filed a patent for a jet engine in Germany, based on his diploma at Göttingen University, where he was unaware of Whittle's work, or Griffith's or Guillaume's. Ohain had a better reception from German industry and in March 1937 his engine was ready for its first test run at the Heinkel plant in Rostock. A month later, Whittle's design also came into existence and ran for the first time at the British Thomson Houston company in Rugby. Whittle had revived his project as a company, Power Jets, with the backing of industrialists, in 1935. As examples of simultaneous innovation go, this parallel story of Whittle and Ohain, with its almost exact matches in date, is extreme – but the general phenomenon surprisingly common.

The parallels continued. Ohain's jet engine got a Heinkel plane into the air well before Whittle's, the first flight happening on 27 August 1939, just a few days before the invasion of Poland began the Second World War. Whittle's engine got a Gloster plane into the air on 15 May 1941. Both Germany and Britain had jet fighters in combat for the first time in the same month, the Gloster Meteor shortly after 17 July 1944, the Messerschmitt 262 on 25 July, but though they were fast they had little influence on the war, being limited in range. Britain gave America the technology during the war and American jets were also aloft by the end of the war.

Later Whittle, somewhat embittered and without much financial reward, wrote his memoirs as a tale of lonely

genius struggling against official, bureaucratic and corporate resistance, but subsequent historians have revised this
account, finding that the British government and British
industry were actually fairly receptive – at least by their
sluggish standards – to Whittle's ideas, and that the story
of the jet was a much more collective effort than it seems
at first. Indeed, the main design for jet engines today uses
Griffith's axial flow, whereas Whittle used centrifugal flow.
As Andrew Nahum has put it: 'Rather few historians, or
indeed engineers, given a moment to reflect, would now
assert that there would have been no jet engine without
Whittle.'

The same could be said of Ohain. Both were brilliant
pioneers, who affected the course that history took, but the
jet engine would have happened without them. Curiously,
they did not meet until 1966, when Ohain was working for
the US Air Force and Whittle was long retired.

Like radar and the computer, the jet is often thought
to be a product of wartime inventiveness. But, as in those
other cases, the key work was actually done long before
hostilities broke out, both in Britain and in Germany, and it
is impossible to know just how fast the jet would have been
developed and commercialized in an alternative universe in
which the 1940s were prosperous and peaceful.

After the Second World War, the race to improve and
perfect the jet engine for passenger aircraft, as well as
for military planes, was mainly carried out within three
big companies: Pratt and Whitney, General Electric and
Rolls-Royce. The heroic age was over; now it was teams
of engineers doing thousands of experiments and reams of
calculations, incrementally inching up the power and efficiency of jet engines, in turning heat into work, until they

reached today's 40 per cent, compared with just 10 per cent in Ohain's and Whittle's first jets.

Innovation in safety and cost

The truly extraordinary improvement in the safety record of air travel is an example of gradual but pervasive innovation with real impact. In 2017, for the first time, there was no death as a result of a commercial passenger jet crashing. There were fatal crashes involving cargo planes, private planes and propeller aircraft, but no commercial passenger jets. Yet that year also saw a record 37 million commercial flights. The number of airline accident fatalities in the world has declined steadily from over 1,000 people a year in the 1990s to just 59 in 2017, even as the number of people flying has greatly increased. The general trend remains true despite the two accidents suffered in 2018 in Indonesia (189 fatalities) and 2019 in Ethiopia (157 fatalities) by Boeing 737-MAX 8 planes, caused by computer error. These two exceptional tragedies underlined just how rare such accidents had become and resulted in the grounding of the entire fleet of such planes.

The comparison with half a century ago is even starker. There are now more than ten times as many people in the air at any one time as there were in 1970, yet according to the Aviation Safety Network the number of fatalities was more than ten times greater in the earlier year. In 1970 there were 3,218 fatalities per trillion revenue-passenger-kilometres. In 2018 there were just fifty-nine – a 54-fold decline. In America, you are now at least 700 times more likely to die in a car, per mile travelled, than in a plane.

The decline in air accidents is as steep and impressive as the decline in the cost of microchips as a result of Moore's Law. How has this been achieved? The answer, as with most innovation, is that it happened incrementally as a result of many different people trying many different things. To take an early example, in the 1940s Alphonse Chapanis, tasked with identifying the causes of accidents in the Army Air Corps, noticed that tired pilots were sometimes retracting the landing gear instead of the flaps as they touched down. The controls for the two were identical in shape and next to each other. He recommended changing both the location and the shape of the controls, so that wheel controls looked like wheels and flap controls looked like flaps.

More generally, it is the widespread use of dull, low-tech but vital practices such as 'crew resource management' techniques, and checklists galore, with cross-checking between crew members, and a culture of challenge, that have made the big difference since the 1970s.

In 1992 an Air Inter flight on a modern Airbus 320 crashed into a mountain while coming in to land at Strasbourg airport, killing eighty-seven of ninety-six people on board. The weather was snowy, it was dark, but many more factors contributed to the accident, all of them avoidable. The primary cause was that the crew selected the wrong mode in the flight management and guidance system: 'vertical speed' mode instead of 'flight path angle' mode. This mistake was too easy to make, and too hard to spot. This meant that when they entered the number '33', the aircraft began descending at 3,300 feet per minute instead of at 3.3 degrees, but the display did not make this clear enough. That air traffic control had given the crew a wrong fix, causing confusion on the flight deck, and that the crew were not communicating well or cross-checking with each other, contributed too. Finally, the aircraft was not equipped with a ground-proximity warning system, because it was thought likely to produce too many false alarms in mountainous regions. As cases like this show, there are multiple factors of technology, procedure and psychology that safety designers have to get right in making flight safer. Most crucial of all is learning from mistakes such as this accident, by openly and transparently sharing the results of accident investigations all around the world. The astonishing safety record of the modern airline industry has been achieved, quite literally, by trial and error. Its methods have since been emulated in other walks of life such as surgery and offshore oil and gas exploration.

This improvement in safety has happened in an era of deregulation and falling prices. Far from leading to cut corners and risk taking, the great democratization of the airline industry over the past half-century, with its fast turnarounds, no-frills service and cheap tickets, has coincided with a safety revolution. Herb Kelleher, who died in 2019 at the age of eighty-seven, is a strong candidate for the most prominent hero of the budget-airline revolution. He founded Southwest airlines in 1967 in an era when commercial flight was run as a cartel of government-sponsored and often nationalized carriers. Flights between states within America were determined entirely by the government, with airlines taking instruction from the Civil Aeronautics Board when setting prices and deciding on routes.

Kelleher therefore decided that his airline could not leave Texas initially. Even so, three existing airlines immediately got a restraining order to prevent him flying. He lost court case after court case against this cartel until the Texas Supreme Court ruled unanimously in his favour. Even then the legal battles persisted, but Kelleher was a lawyer and knew how to fight them. As the author Jibran Khan related, two airlines, Braniff and Texas International, lodged complaints against him with the federal Civil Aeronautics Board. But Kelleher argued his case in court and won, the board dismissing the objections. The two airlines found another judge, who had ruled against Southwest a few years earlier in another case, and got another injunction. The Texas supreme court went into emergency session and overturned the injunction. In 1977 Braniff and Texas International would be indicted for conspiring to monopolize the industry.

Southwest took to the skies at last in 1971, and by 1973 it was profitable despite offering low fares. It remains so to this day, an unmatched record in an industry scarred by bankruptcies and mergers. Kelleher's innovations included the simple idea that flight attendants should be encouraged to crack jokes, and when a flight is ready to take off but the food has not arrived, to take a straw poll among the passengers as to whether to wait for the food (the vote usually goes for not waiting).

In 1974 the government set ticket prices on airlines, and the minimum for a standard-class ticket from New York to Los Angeles was more than $1,550, in today's dollars. Today it is a fraction of that. Many imitators of Kelleher's initiative have since followed the same path of cost cutting, with varying degrees of success, from Freddie Laker to Michael O'Leary to Bjørn Kjos, the founder of Norwegian Air. These are the true transport innovators of today, the heirs to Stephenson and Ford.

Pause in awe at what innovation does. For the entire history of humanity before the 1820s, nobody had travelled faster than a galloping horse, certainly not with a heavy cargo; yet in the 1820s suddenly, without an animal in sight, just a pile of minerals, a fire and a little water, hundreds of people and tons of stuff are flying along at breakneck speed. The simplest ingredients – which had always been there – can produce the most improbable outcome if combined in ingenious ways. Early in the following century, people are taking to the air, or piloting their own carriages along roads, once again just through the rearrangement of molecules and atoms in patterns far from thermodynamic equilibrium.

4

Food

No potatoes, no popery!
The mob, 1765

The tasty tuber

The potato was once an innovation in the Old World, having been brought back from the Andes by conquistadors. It provides a neat case history of the ease, and difficulty, with which new ideas and products diffuse through society.

Potatoes are the most productive major food plant, yielding three times as much energy per acre as grain. They were domesticated about 8,000 years ago in the high Andes, above 3,000 metres, from a wild plant with hard and toxic tubers. Quite how and why people managed to select a nutritious plant from such a dangerous ancestor remains shrouded in the mists of time, but it probably happened somewhere near Lake Titicaca. Francisco Pizarro and his band of conquerors encountered the potato and ate it while decapitating and looting the Inca Empire in the 1530s. But the conquistadors' emphasis was on taking their familiar

Old World crops and animals to the New World, more than vice versa, so it was at least three decades before the potato appeared on the eastern side of the Atlantic. Maize, tomatoes and tobacco got back to the Old World much quicker.

The first definite account of potatoes grown on the eastern side of the Atlantic comes from the Canary Islands, where the archives of the public notary in Las Palmas de Gran Canaria contains a list of the goods shipped by Juan de Molina to his brother, Luis de Quesada, in Antwerp on 28 November 1567: 'three medium-sized barrels [which] you state contain potatoes and oranges and green lemons'.

Slow to arrive, the potato was slow to catch on in Europe. Against it was a combination of practice and prejudice. Being from the tropics, potatoes were adapted to twelve-hour days and would not produce tubers in the longer days of European summers, so it was autumn before they 'fruited' and then disappointingly. It was probably in the Canaries that selection and breeding gradually solved this problem.

As for prejudice, clergymen forbade their parishioners from eating potatoes in England as late as the early eighteenth century, for the magnificently stupid reason that they are not mentioned in the Bible. Somehow, probably with an Irish flavour to the argument, the English turned this into a belief that potatoes were Roman Catholic agents: in Lewes, in Sussex, the crowd shouted: 'No potatoes, no popery!' during an election in 1765. Yet in rainy Lancashire and in Ireland, the potato's ability to yield reliable harvests, even in wet years when the grain crop rotted, proved irresistible. In 1664 one John Forster wrote a pamphlet urging the king to make money from royalties on potato growing. He offered, in the title alone:

> a Sure and Easie Remedy Against all Succeeding Dear Years; by a Plantation of the Roots called POTATOES, whereof (with the addition of wheat flour) Excellent, Good and Wholesome Bread may be Made, Every Year, Eight or Nine Months Together, for Half the Charge as formerly.

The potato also had to overcome a loopy doctrine taught by what passed for intellectuals in those days that plants were good at curing the diseases they most resembled. Walnuts, looking like brains, were good for mental illness. (God liked to drop hints.) This idea was started in the 1500s by the alchemist and astrologer Paracelsus (real name Theophrastus von Hohenheim) and credulously repeated by various herbalists in the sixteenth century. Since potatoes supposedly resembled fingers with leprosy, but since leprosy was very rare, people were somehow induced to think that potatoes might cause leprosy. In 1748 the French parliament, in an early example of the precautionary principle later to inhibit the use of genetic modification, banned the growing of potatoes for human food – just in case they did cause leprosy.

Deterred by such fears, continental Europeans and North Americans only slowly came to like growing and eating potatoes. Indeed, the potato may have spread more rapidly in India and China in the 1600s than in Europe. It took to the Himalayas especially well, being reminded no doubt of the Andes. In eighteenth-century continental Europe the potato as a field crop – rather than a garden delicacy – appears to have spread south from the coast of what is now Belgium and north-west from Alsace, with Luxembourg getting into the habit of growing potatoes in the 1760s and most of Germany by the end of the 1770s. One of the factors that overcame resistance was war. In a world dependent on wheat and barley, invading armies stripped the barns of stored grain and of animals, and trampled or grazed the crops, leaving the population to starve. Potatoes, however, often survived these depredations, being in the ground during the campaigning season and taking too much trouble for soldiers to lift. Farmers who planted potatoes therefore tended to survive better during wars, spreading the habit. As John Reader recounts, the result of Frederick the Great's wars was that the potato, unknown or despised in most of central and eastern Europe in 1700, had by 1800 become an indispensable part of the European diet.

France lagged behind. The French took nervous note of the rich and fattening diet the Prussians were now on, and the demographic threat they thus posed. Here at last, late in the story, we get a glimpse of an individual as potato innovator, at least according to legend. Antoine-Augustin Parmentier was an apothecary working with the French army who rather carelessly managed to get himself captured no fewer than five times by the Prussians during the Seven Years War. They fed him on nothing but potatoes, and he was surprised to see himself growing plump and healthy on the diet. On his return to France in 1763 he devoted himself to proselytizing the benefits of the potato as the solution to France's repeated famines. With grain prices high after poor harvests, he was pushing at an open door.

Parmentier was a bit of a showman and he devised a series of publicity stunts to get his message across. He got the attention of the queen, Marie Antoinette, and persuaded her to wear potato flowers in her hair, supposedly after a contrived encounter in the gardens of Versailles. He planted a field of potatoes on the outskirts of Paris and posted guards to protect it, knowing that the presence of the guards would itself advertise the value of the crop, and attract hungry thieves at night, when the guards were mysteriously absent. He gave dinners of potato cuisine to people of influence, including Benjamin Franklin. But he was also scientific in his approach. His 'Examen chimique des pommes de terre', published in 1773 (a year after the parliament had repealed the ban on potatoes), praised the nutrient contents of potatoes. In 1789, on the brink of revolution and against a background of widespread hunger, the king ordered Parmentier to produce another treatise on 'the culture and use of the potato' as well as other roots. Not that this saved the king's head. It was left to the revolutionaries to reap the full benefits, growing potatoes in the gardens of the Tuileries, preventing mass starvation during the First Commune.

In Ireland the potato fuelled a population explosion that soon threatened to be a Malthusian disaster. The rapidly growing population of the early nineteenth century tilled every acre they could find, reaching a higher density of people per acre than anywhere in Europe, with large families managing to survive to adulthood but in increasingly desperate poverty as land was divided among offspring. As Cecil Woodham Smith details in her book *The Great Hunger*, in the early 1800s:

no fewer than 114 Commissions and 61 Special Committees were instructed to report on the state of Ireland,

and without exception their finding prophesied disaster; Ireland was on the verge of starvation, her population rapidly increasing, three-quarters of her labourers unemployed, housing conditions appalling and the standards of living unbelievably low.

The crash came in 1845 when a parasitic blight fungus (*Phytophthora infestans*) that the potato plant had left behind in the Andes reached Ireland via the United States. That September throughout Ireland the potato crops rotted in the fields both above and below ground. Even stored potatoes turned black and putrid. Within a few years, a million people had died of starvation, malnutrition and disease, and at least another million had emigrated. The Irish population, which had reached over eight million, plunged and has still not returned to the level it was in 1840. Similar if less severe famines caused by blight drove Norwegians, Danes and Germans across the Atlantic.

Today, the potato is experiencing a new wave of innovation. The invention of synthetic fungicides in the 1960s enabled potato farmers to keep the disease at bay, but only by spraying their crops on an almost weekly basis, or up to fifteen times in a season. Then in 2017 the United States approved the release of new potato varieties that are resistant to blight. These had been developed by the J. R. Simplot company in Idaho by genetic modification, specifically through the introduction of a disease-resistance gene from a variety of potato found in Argentina. The new variety requires little if any spraying. Other blight-resistant varieties developed by gene editing are also coming to the market.

How fertilizer fed the world

Fritz Haber's 1908 discovery of how to fix nitrogen from the air to make ammonia, by reacting it with hydrogen in the presence of a catalyst under pressure, stands as one of the key innovations of all time. Not just for the immense impact it had on feeding the world and defeating famine, nor just for the less benign effect of making the manufacture of explosives much easier, but because it was an unusual example of solving an apparently impossible problem. How to make useful compounds of nitrogen out of the air, which is made largely of molecular nitrogen, was a challenge that everybody could see was well worth solving. But by the time Haber did it, most people had concluded that it was as hard to solve as the alchemist's dream of turning lead into gold and might never yield. This is an example of an innovation that the world demanded and got.

That nitrogen was a limiting nutrient in the growing of crops had been known, at least vaguely, for centuries. It led farmers to beg, borrow and steal any source of manure, urea or urine they could find. Try as they might, though, they struggled to apply enough nitrogen to enable their crops to realize their full potential. The best way involved not just manure from cattle, pigs and people, but 'break crops' of peas and beans. These legumes thrived without manure, because they could somehow fix nitrogen from the air, and left the soil enriched for the next year's crop as well. If they could do it, why not a factory?

The science that explained this hunger for nitrogen came much later, with the discovery that every building block in a protein or DNA molecule must contain several nitrogen atoms, and that though the air consisted mostly of nitrogen atoms they were bound together in tight pairs, triple covalent bonds between each pair of atoms. Vast energy was needed to break these bonds and make nitrogen useful. In the tropics, frequent lightning strikes provided such energy, keeping the land a little more fertile, while in paddy-rice agriculture, algae and other plants fix nitrogen from the air to replenish the soil. Temperate farms, growing crops such as wheat, were very often nitrogen-limited, if not nitrogen-starved.

In 1843 a field called Broadbalk was set aside at Rothamsted in Hertfordshire to demonstrate the effect of fertilizer. One strip of the field has been planted every year since then with winter wheat and with no fertilizer of any kind. It became a tired and desolate sight, yielding less and less grain, till by 1925 it was able to produce less than half a tonne from every hectare, a small fraction of that which could be harvested from a part of the field that received farmyard manure or nitrate fertilizer. After 1925 fallow was introduced into the rotation, so that the land could recover some nitrogen from wild clover every other year. Yield rose on the untreated strip but only to modest levels. The lesson for humanity is obvious: without a continuous input of nitrogen, from crops grown elsewhere or at other times and perhaps fed to cattle or people first and turned into manure, farming cannot feed people sustainably.

During the nineteenth century this did not matter all that much. The plough marched west into the prairies, east into the steppes and south into the pampas and the outback, breaking virgin soil that had been denuded of its wild grazing herds and its native people, and unleashing its fertile potential. More land fed more mouths. That the land soon became exhausted unless replenished by manure or clover mattered, but there was always new land to break. Westward Ho!

It did not help that there was a competing demand for nitrogen. Kings and conquerors also coveted ionized nitrogen (not that they knew it as such), with which to make gunpowder and wage war. In 1626, for instance, King Charles I of England ordered his subjects to 'carefully and constantly keep and preserve in some convenient vessels or receptacles fit for the purpose, all the urine of man during the whole year, and all the stale of beasts which they can save', and with this to make saltpetre, the basic ingredient of gunpowder. Farmers all over the world were forced to make saltpetre from manure and pay it as a tax, to support the monopoly on violence claimed by their rulers, thus depriving their fields of a source of fertilizer. One of the motives for the British conquest of Bengal was to gain access to the rich saltpetre deposits at the mouth of the Ganges.

In the early 1800s the world stumbled upon a huge mother-lode of fixed nitrogen, combined with two other elements vital to plants, phosphorus and potassium. Off the coast of Peru were some small islands in a sea rich in fish. This combination of circumstances attracted millions of breeding birds, mainly shags and boobies. Since it almost never rained here to wash the islands clean, their rich droppings had accumulated, century after century, till there was a grey guano soil hundreds of feet deep, steeped in urea, ammonium, phosphate and potassium. Perfect for enriching yields of farms. Over the middle decades of the nineteenth century, millions of tonnes of guano were mined in horrific conditions, by mainly Chinese indentured labourers who were little more than slaves, to satisfy the needs of farmers in Britain and other parts of Europe. Ships queued for months to await the chance to load the dusty and foul-smelling cargo.

Desperate to get access to guano, the American Congress passed an Act saying that any American who found a guano island in the Pacific could claim it for the United States – which is why so many mid-Pacific atolls belong to America today. Few islands proved as rich as the Chinchas off Peru. The Namibian coast had a similar combination of rich sea and dry desert air, and a Liverpool merchant opened a guano mine here on Ichaboe island in 1843. By 1845, he was filling up to 400 ships, steadily reducing the height of the island and fighting pitched battles with rival miners. But Ichaboe and the Chinchas soon began to run out of guano. Today, cormorants, boobies and penguins are back on the islands, slowly rebuilding the guano.

The guano boom made great fortunes, but by the 1870s it was over. It was succeeded by a boom in Chilean saltpetre, or salitre, a rich nitrate salt that could be made by boiling caliche, a mineral found in abundance in the Atacama desert, the result of desiccated ancient seas uplifted into the mountains and left undissolved by the extreme dryness of the climate. Though the mines and refineries were mostly in Peru and Bolivia it was Chileans who worked them, and in 1879 Chile declared war and captured the key provinces, cutting Bolivia off from the sea and amputating part of Peru. By 1900 Chile was producing two-thirds of the world's fertilizer, and much of its explosive. But the best deposits of Chilean nitrate, too, were soon showing signs of running out.

It is against this background that a speech by a famous British chemist suddenly caught the world's attention. Sir William Crookes, a wealthy and independent scientist and spiritualist, famous for discovering thallium, isolating helium and inventing the cathode ray tube, was elected president of the British Association for the Advancement of Science in 1898. This was a year-long job that brought with it the obligation at the end to make a formal speech and say something profound. He chose to speak about the 'wheat problem', namely the looming probability that the world would be starving by 1930 unless a way could be made to synthesize nitrogen fertilizer to replace Chilean nitrate, wheat being then by far the largest crop in the world.

Crookes's warning was noticed especially in Germany, a country that was using larger and larger sailing ships to import more Chilean nitrate than any other nation, in order to support a growing population. As Britain went to war with South African Boers, of Dutch and German extraction, the year after Crookes's speech, a distinguished German chemist named Wilhelm Ostwald began to wonder: what if there were a war and Britain's Royal Navy cut off the Chilean trade as a way of depriving Germany of the raw material with which to make gunpowder and fertilizer? Ostwald joined the race to fix nitrogen from the air, but instead of using electricity, as most were trying, he tried chemical catalysts, especially iron. In 1900 he thought he had succeeded in making ammonium, but Carl Bosch, employed by the BASF chemical company to check before they bought his patents, discovered that it was a mirage. The ammonia was a contaminant of the iron, derived from iron nitride. Ostwald retired hurt.

Enter Fritz Haber. An ambitious, prickly, restless genius, sensitive about his Jewish background and suspecting (rightly) that anti-Semitic discrimination was holding him back from the glittering prizes that were his due, but also fiercely nationalistic on behalf of imperial Germany, Haber too saw the fixing of nitrogen as a golden goal. It would effectively make 'bread from air' in the arresting phrase of Haber's modern biographer, Thomas Hager. In 1907 Haber had a feud with Walther Nernst, Ostwald's protégé, when he first claimed to have made ammonia in small quantities using heat and a catalyst. Nernst said Haber could not possibly have made even as little as he claimed. Furious, Haber returned to the laboratory determined to prove Nernst wrong, but also having picked up a hint, from Nernst, that using very high pressure might work. He soon found that the higher the pressure, the lower the temperature at which the reaction worked. This was crucial because very high temperature caused the ammonia to self-destruct almost as soon as it formed. Haber's assistant, Robert Le Rossignol, gradually figured out, step by step, how to hold the ingredients together at high pressure inside a chamber drilled out of solid quartz. 'There was no single moment of breakthrough, just a number of small improvements and incremental advances,' writes Hager.

It was at this point that Haber approached BASF, the giant chemical company that had grown rich from making synthetic indigo dye and was looking for a second act. BASF was determined to crack nitrogen fixation, but thought electricity was the way to go. It invested in Haber's idea mainly as a fallback. It gave him a laboratory, a big budget, 10 per cent of any sales and the chance to stay on at Karlsruhe University. With BASF's money and expertise Haber and Le Rossignol were able to push their experiments above 100 atmospheres of pressure, the equivalent of a mile beneath the surface of the sea, and to drop the temperature from over 1,000°C to 600°C.

But results were disappointingly far from being commercially viable, so Haber began to try different catalysts. Like Edison seeking the right material for a filament in a light bulb, Haber cast about for different metals almost at random, and indeed it was from lighting filaments that he eventually stumbled upon osmium, a dense, shiny blue-black metal element usually found alongside platinum and first described in 1804. In March 1909 Haber watched liquid ammonia dribble out of the apparatus at the second test of an osmium catalyst. He had no idea why osmium worked, but it did.

He immediately proposed to BASF that they scale up his idea. The company was sceptical: osmium was rare and expensive, while a manufacturing plant running at 100 atmospheres of pressure without exploding was impossible to imagine, let alone build at a reasonable price. But Carl Bosch, the man who had exposed Ostwald's failure nine years before, and was now in charge of nitrogen research at BASF, argued for giving it a go, mainly because he was out of other ideas.

Over the next few years Bosch turned Haber's invention into a practical innovation, solving problem after problem in the quest to build a factory, rather than a toy, to produce ammonia by the tonne rather than by the teaspoon, and to do so more cheaply than it could be shipped from Chile. He first bought up almost the world's entire supply of osmium, a few hundred kilograms, but it was not enough. Haber discovered that uranium worked also, though not as well, but it was not much cheaper or more abundant. So Bosch set up a plant in which new catalysts could be tested, and new designs for containing the high-pressure ingredients at the same time. It was kept behind a strong wall, so that it could explode without killing people. Eventually, Bosch's assistant, Alwin Mittasch, went back to pure iron, then iron compounds, and found one sample of magnetite from Sweden that got good results. Some impurity in the magnetite made the iron into a good catalyst. By the end of 1909, they had settled on a mixture of iron, aluminium and calcium. It worked as well as osmium, but was far cheaper. Mittasch went on searching for still better catalysts, testing over 20,000 different materials, but he never improved on the iron mixture. BASF made Haber keep quiet about the catalyst, though it allowed him to announce the breakthrough with osmium in 1910, giving the firm tacit knowledge that kept it in the lead.

Vast challenges remained: how to purify nitrogen from the air; how to make enough hydrogen out of steam exposed to hot coke without including carbon monoxide in the gas too; how to achieve unprecedentedly high pressures; how to contain such pressures at red-hot temperatures; how to feed in the gases and extract the ammonia. The team grew into the biggest group of scientists and engineers before the Manhattan Project. The Haber–Bosch story, like so many about innovation, is often told as one of brilliant insight by academic (Haber) followed by inevitable application by businessman (Bosch), but this is wrong. Far more ingenuity was needed during Bosch's perspiration than during Haber's inspiration. As Hager recounts, none of these challenges could be overcome without access to the ideas being developed in other industries, a fine example of how innovation thrives in an ecosystem of innovation:

Bosch's teams looked for design hints in locomotive engines, gasoline engines, and the new engine that Rudolf

Diesel had invented. Bosch and his engineers met with men from the German steel industry, learned about the Bessemer process for making steel, talked with Krupps representatives about cannon designs and new advances in metallurgy. He set teams to work designing quick-acting valves, self-closing valves, slide valves; pumps reciprocating and circulating, large and small; temperature monitors of all sorts and sizes; pressure balances; density recorders; trip alarms; colorimeters; high-pressure pipe fittings. Everything had to be rugged, leakproof, functional at high temperature and under enormous pressure. The ovens had the potential of exploding like small bombs; Bosch wanted to make sure they could be carefully monitored and quickly shut down if something started to go wrong. He wanted perfect reliability and lightning speed. He wanted a machine that combined the strength of a sumo wrestler, the speed of a sprinter, and the grace of a ballerina.

For six months Bosch was held up by an apparently insuperable problem: that hydrogen infiltrated the steel walls of the ovens and weakened them, leading them to explode after a few days. He tried different alloys, but nothing helped. Only by rethinking his whole approach, and using a sacrificial layer of weaker steel within, and boring small holes to exhaust the hydrogen from between the two layers was he able to control this problem. By 1911 he had prototypes running continuously and producing ammonia cheaply – so long as you wrote off the costs of developing the system.

As so often, intellectual property now got in the way. The rival firm Hoechst, advised by Ostwald, challenged Haber's patent on the production of ammonia with heat and pressure, arguing that Nernst had started the whole idea under Ostwald's direction. BASF, facing ruin, simply bought off Nernst with a lucrative five-year contract in exchange for his testifying on their side in court.

The company's huge factory at Oppau began producing ammonia in late 1913, just in time for the First World War. Germany had a store of Chilean nitrate to use for making explosive that it thought would last a short war, and it captured more when Antwerp fell into German hands. But when the war bogged down in trench stalemate, and the Royal Navy sank a German fleet that had been blockading the Chilean nitrate trade, in a battle off the Falkland Islands, Germany faced the prospect of running out of fixed nitrogen to make explosives for guns and fertilizer for fields, just as Ostwald had feared. In the short run it began to make small amounts of nitrate using electricity in the expensive Cyanamid process, using electricity and calcium carbide.

Then in September 1914 Bosch made the famous 'saltpetre promise' that he could convert the Oppau plant so that it turned ammonia into nitrate, using a newly discovered iron-bismuth catalyst. He built an even bigger plant at Leuna, producing huge quantities of nitrate and thus probably prolonging the war. Haber, in the meantime, had invented gas warfare, personally presiding over the first chlorine attack at Ypres in March 1915.

After the Great War, the Haber–Bosch process was used throughout the world to fix nitrogen on a grand scale. The process became steadily more efficient, especially once natural gas was substituted for coal as the source of energy and hydrogen. Today, ammonia plants use about one-third as much energy to make a tonne of ammonia as they did in Bosch's day. About 1 per cent of global energy is used in nitrogen fixation, and that provides about half of all fixed nitrogen atoms in the average human being's food. It was synthetic fertilizer that enabled Europe, the Americas, China and India to escape mass starvation and consign famine largely to the history books: the annual death rate from famine in the 1960s was 100 times greater than in the 2010s. The so called Green Revolution of the 1960s and 1970s was about new varieties of crop, but the key feature of these new varieties was that they could absorb more nitrogen and yield more food without collapsing (see next section). If Haber and Bosch had not achieved their near-impossible innovation, the world would have ploughed every possible acre, felled every forest and drained every wetland, yet would be teetering on the brink of starvation, just as William Crookes had forecast.

Yet, as I write this, it is possible to glimpse a future in which the Haber–Bosch process is redundant. In 1988 two Brazilian scientists, Joanna Döbereiner and Vladimir Cavalcante, noticed something peculiar. Some fields of sugar cane were producing consistent yields without having received any fertilizer for decades. They searched inside the plant tissues and found a bacterium, *Gluconacetobacter diazotrophicus*, that was fixing nitrogen from the air. This ability is found in legumes such as peas and beans, thanks to a symbiosis between the plants and bacteria that live in special nodules on the roots. But all attempts to persuade crops like maize and wheat to emulate this legume habit had so far failed. Perhaps this new bacterium, which lived inside the plant and did not need special nodules, might do better. A sample of the bacteria reached Professor Ted Cocking of Nottingham University, and he soon persuaded the bacterium to live inside the actual cells of various species of plant. Remarkable improvements in the yield and protein content of maize, wheat and rice were soon being shown in field trials. In 2018 the company Cocking founded with David Dent, Azotic, announced that it was going to market the bacteria as a seed dressing to American farmers. If this simple fix succeeds, it may prove possible to feed the world without ammonia made in factories.

Dwarfing genes from Japan

Around the time that Bosch was perfecting the fixation of air, on the other side of the world, a plant breeder was pursuing a different innovation that would prove vital to the application of Bosch's product.

In 1917 at the Central Agricultural Experiment Station in Nishigahara, near Tokyo, somebody, it is not clear who, decided to cross two varieties of wheat. One was called 'glassy Fultz' and it was derived from a wheat variety imported from the United States in 1892. The other was a native Japanese variety of dwarf stature, known as Daruma. The resulting wheat, Fultz–Daruma, was then crossed in 1924 with another American variety called Turkey Red. Samples of this wheat were grown and self-crossed before being tested at an agricultural research station at Iwate in north-east Japan. The best plants seemed to retain the short stature of Daruma and the high yield of Turkey Red. The station head, Gonjirô Inazuka, selected the most promising lines and in 1935 began to market a true-breeding new wheat variety under the name Nôrin-10. Local farmers began commercially growing dwarf wheat for the first time.

Ten years later, in the aftermath of war, there arrived in Japan an agronomist and wheat-breeding expert from

Kansas by the name of Cecil Salmon. He was serving on the staff of General Douglas MacArthur, the de facto ruler of Japan. Salmon was intrigued by the dwarf wheats he saw at Morioka Agriculture Research Station in Honshu and sent sixteen samples back to the small-grains collection in the United States. One of these was Inazuka's Nôrin-10.

Meanwhile, a third wheat breeder, Orville Vogel at Washington State University in Pullman, was wrestling with a problem caused by Haber–Bosch's nitrate fertilizer. Applied to fields it caused wheat plants to grow thick and tall. This meant that as soon as the wind blew and the rain fell, the ripening wheat crop would tend to collapse under its own weight, or 'lodge', then lie flat and rotting on the ground. Salmon's seeds from Japan came to his rescue, via a fourth breeder named Burton Bayles. Vogel takes up the story:

> Being aware of our lodging problems, B. B. Bayles sent a collection of semidwarf wheats for preliminary observations at Pullman in 1949. From these, Norin 10 was selected to be crossed with Brevor which at that time was considered to be the most lodging resistant high yielding variety with short straw.

Perhaps, reasoned Vogel, even shorter straw would make the wheat less likely to topple, save it from lodging and allow it to adapt to the new fertilizer. Sure enough, some of his new Nôrin-10 crosses, especially the cross with Brevor, proved capable of staying upright while yielding 'real good' – as Vogel's notebook records. The only problem was that they were susceptible to local diseases, so Vogel continued experimenting in search of a less susceptible line before putting it on the market.

A fifth wheat breeder now got to hear of Vogel's experiments and asked him for some samples. This was Norman Borlaug, a Minnesotan descended from refugees who left Norway during a potato famine. After an aborted career as a forester, Borlaug was working for the Rockefeller Foundation in Mexico, where his aim was to find varieties of wheat resistant to rust fungus and with good yields.

Borlaug and his team were making good progress. At first, even though yields were excellent, no Mexican farmer trusted the new varieties. Eventually in 1949 Borlaug persuaded a few to plant them, and to use fertilizer on them. The news of their higher yields began to spread. Farmers found they could double their yields and double their incomes. By 1951 the wheat crops were swelling all across Mexico. By 1952 Borlaug's wheats dominated the country's wheat acreage and the entire country's wheat production had doubled.

Soon Borlaug, like Vogel, became distracted by the lodging problem. He searched the entire collection of American wheat varieties for one that could resist falling over, without success. On a trip to Argentina, he found himself talking over a drink to the same American government wheat breeder, Burton Bayles, who had sent the Nôrin seeds to Vogel. He asked Bayles if he knew of any short-strawed wheats that resisted lodging. Bayles told him about Nôrin-10 and suggested he contact Vogel. Borlaug wrote to Vogel, who sent him both pure Nôrin-10 seeds and Nôrin-Brevor hybrids. Borlaug began crossing them with his Mexican wheats. He got spectacular results:

Not only was dwarfness of stature introduced into the crosses from the Norin 10 derivatives, but also a number

of other genes had been introduced, which increased
the number of fertile florets per spikelet, the number of
spikelets per head and the number of tillers per plant.

Like Vogel, Borlaug found the new varieties suffered from
rust infection. But he had an advantage over the Washington team: he was growing the wheats at two different
locations, at very different altitudes, meaning that the low-
elevation, irrigated wheat in the northern Sonora Valley
could be harvested before the high-elevation crop in the
central highlands was planted. He was thus able to get two
breeding seasons in a year. He tested tens of thousands of
varieties for rust resistance. By 1962 Borlaug had a commercially viable variety to offer Mexican farmers with
short straw, massive yields if well fertilized, little lodging
and good rust resistance.

Mexico was only one country. Enter a sixth wheat
breeder, by the name of Manzoor Bajwa, of Pakistan.
Bajwa met Borlaug when the latter came to Pakistan in
1960; he immediately applied to go and work with him
in Mexico. There among the crosses he identified a line of
wheat that was short-strawed and rust-resistant to test in
the Indus Valley. The new variety caught the attention of
the Minister of Agriculture in West Pakistan, Malik Khuda
Bakhsh Bucha. But the Pakistani scientific establishment
was scornful, telling Borlaug and Bajwa that the Mexican
wheats were unsuited to Pakistan, susceptible to disease,
dependent on fertilizer, which only made weeds grow; or
more fancifully that the genes in the new varieties might
sterilize cattle or poison Muslims and were a CIA plot to
make the country dependent on American technology. So
progress stalled.

Across the border in India, a seventh wheat geneticist in this story, Momkombu Sambasivan Swaminathan, had also taken notice. He invited Borlaug to India in 1963 to help persuade his government to embark on a crash programme of wheat improvement. It was uphill work. As Borlaug later said:

> When I asked about the need to modernize agriculture, both scientists and administrators typically replied, 'Poverty is the farmers' lot; they are used to it.' I was informed that the farmers were proud of their lowly status, and was assured that they wanted no change. After my own experiences in Iowa and Mexico I didn't believe a word of it.

The Indian bureaucrats were adamant that Mexican wheats should not even be allowed in the country, let alone encouraged. The biologists warned of devastation and disease if the wheats failed. The social scientists warned of 'irreversible social tensions' and riots if the wheats succeeded – and caused some farmers to make more money than others. Thus do innovation's opponents seek any argument, however absurd, to defend the status quo.

Yet India should have been desperate for new ways to feed its growing population; hunger and malnutrition were widespread. Towards the end of the 1960s, after poor monsoons led to famine, Western experts began to write off India as impossible to feed. The ecologist Paul Ehrlich forecast famines 'of unbelievable proportions' by 1975; another famous environmentalist, Garret Hardin, said feeding India was like letting survivors of a shipwreck climb aboard an overloaded lifeboat; the chief organizer of

Earth Day, in 1970, said it was 'already too late to avoid mass starvation'; a pair of brothers, William and Paul Paddock, one an agronomist and the other a Foreign Service official, wrote a best-seller called *Famine 1975!*, arguing for abandoning those countries, like India, that were 'so hopelessly headed for or in the grip of famine (whether because of overpopulation, agricultural insufficiency, or political ineptness) that our aid will be a waste; these "can't-be-saved nations" will be ignored and left to their fate'. Never have gloomy and callous forecasts been so rapidly proved wrong. Both India and Pakistan would be self-sufficient in grain within a decade thanks to dwarf wheat.

In 1965, with determined support from their ministers of agriculture, India ordered 200 tonnes and Pakistan 250 tons of Borlaug's Mexican wheat to plant as seed. Fulfilling the order was a nightmare for Borlaug, as the shipment was held up at the American border on the way to Los Angeles, delayed by the riots in the Watts area of the city and then arrived in Bombay and Karachi in the middle of a brief outbreak of war between the two countries. But the grain reached its destinations just in time for planting and the harvest was promising. Over the next few years, inch by inch Borlaug won over his critics and Pakistan especially began to experience remarkable increases in its wheat harvest.

In India, farmers on the ground soon began to see the difference, but the government refused to license the import of enough fertilizer, or the construction of fertilizer plants by foreign firms, to make the new crops reach their potential. Borlaug's long campaign culminated in a stormy meeting on 31 March 1967 with the deputy prime minister and head of planning, Ashok Mehta. Borlaug decided to throw caution to the winds. In the midst of an argument, he yelled:

> Tear up those five-year plans. Start again and multiply everything for farm support three or four times. Increase your fertilizer, increase your support prices, increase your loan funds. Then you will be closer to what is needed to keep India from starving. Imagine your country free of famine . . . it is within your grasp!

Mehta listened. India doubled its wheat harvest in just six years. There was so much grain there was nowhere to store it. In his acceptance speech on being awarded the Nobel Peace Prize in 1970, Norman Borlaug said that 'man can and must prevent the tragedy of famine in the future instead of merely trying with pious regret to salvage the human wreckage of the famine, as he has so often done in the past.'

This fifty-year story of how dwarfing genes were first found in Japan, cross-bred in Washington, adapted in Mexico and then introduced against fierce opposition in India and Pakistan is one of the most miraculous in the history of humankind. Thanks to Inazuka–Borlaug genetic varieties, and Haber–Bosch nitrogen fertilizer, India not only fed itself, proving the forecasts of worsening starvation wrong, but became an exporter. The dwarfing versions of genes in Nôrin-10 (which turned out to be two mutations known as Rht1 and Rht2 that make the plant less responsive to growth hormones) thus changed the world, in combination with fertilizer fixed from the air. Rice quickly followed suit with its own dwarf varieties and higher yields; so did other crops. A determined campaign to blame this Green

Revolution for various environmental and social problems in the country, such as farmer suicides, proved to be fake news: Indian farmers are actually less likely to commit suicide than average Indians.

Insect nemesis

In 1901 a Japanese biologist named Ishiwata Shigetane began looking into the cause of a lethal disease of silkworms called sotto, or sudden-collapse disease, which had economic implications for the nationally important silk industry. He quickly identified a bacterium as the cause. Little did he realize that nearly a century later his discovery would lead to a vital innovation that would transform farming practice and make it more environmentally friendly as well as more productive: insect-resistant crops.

The same bacterium was rediscovered and named by a German researcher in 1909. Ernst Berliner was studying flour moths at the Research Institute for Cereal Processing in Berlin. A shipment of flour from a mill in Thuringia contained diseased caterpillars, and the disease quickly spread to the flour moths being bred in the laboratory. Berliner isolated the bacterium behind the infection and named it *Bacillus thuringiensis*. It turned out to be the same creature that had been killing the Japanese silkworms. Bt, as it came to be known, possessed an ability to kill the caterpillars of any moth or butterfly because of a gene for producing a crystallized protein that was lethal to such insects. It latches on to receptors in their gut walls and turns those walls porous.

By the 1930s, in France, it had become possible to buy Bt in the form of bacterial spores, as a living insecticide

known as Sporine. It remains on the market today under
the labels Dipel, Thuricide or Natural Guard, and is mainly
used by organic farmers and gardeners, because it is not a
product of the chemical industry but an example of biological control. It has repeatedly been shown to be harmless
to people, the crystal being destroyed by stomach acid in
mammals and being unable to fit mammalian receptors
anyway. Varieties of the bacterium that can kill flies and
beetles were added to the range of products in 1977 and
1983 respectively.

But Bt, though useful in a greenhouse, is not a very cost-
effective spray for farmers, being expensive and patchy
in its results, easily destroyed by sunlight or washed off
by rain. It also often fails to reach insects living inside the
plants, such as cotton bollworms or corn stem borers.

This is where a Belgian biochemist enters the story. Marc
Van Montagu was born in Ghent in 1933 at the height of
the Great Depression. His family lived in poverty and his
mother died giving birth to him. None of his parents or
siblings had finished school, but an uncle was a teacher and
insisted he not only stay at school but go to university too.
He became an expert on the biochemistry of nucleic acids
and then in 1974, together with his colleague Jeff Schell,
made a key discovery – the 'tumour-inducing (Ti) plasmid'.
This was a small, circular chromosome inside a bacterium
called *Agrobacterium tumefaciens*, which was known to
have the strange property of inducing tumours in plants –
known as crown galls – yet not inhabiting those tumours
itself.

Three years later, Van Montagu was narrowly beaten by
Mary Dell Chilton of Washington University in St Louis to
the discovery that the Ti plasmid stitches some of its DNA

into the plant's own DNA as part of the infection. Given
that, a few years before, tools had been developed to insert
genes from animals or plants into bacteria – for example, to
make human insulin for diabetics – now the reverse became
feasible: to insert bacterial genes into plants. Within six
years, in an example of simultaneous invention, teams led
by Van Montagu, Chilton and Robert Fraley of Monsanto
all turned this insight into an invention, by showing that
Agrobacterium could be manipulated to insert any gene
into a plant by removing the tumour-inducing gene from
the plasmid and replacing it with a gene from a different organism. The result was a healthy plant with a new
gene. Agricultural biotechnology was born. It was using
Ti plasmids that scientists would go on to create many
genetically modified crops, including herbicide-tolerant
maize and soybeans, and eventually virus-resistant papaya
and vitamin-enriched 'golden' rice.

Van Montagu set up a company, Plant Genetic Sciences,
to develop the technology. One of the first candidate genes
his colleagues came up with to insert into a plant was the
protein from Bt that kills insects, since it was already popular with organic farmers and gardeners. In 1987, in the
laboratory, they created a tobacco plant that was normal
in every way except that it included Bt's key gene in its
chromosomes. It proved lethal to tobacco horn worm, a
common pest. Soon the technology was licensed by Monsanto to produce cotton, maize, potatoes and other crops
that were inherently resistant to insects.

Because the insecticidal protein was within the plant, it
would kill caterpillars that bored inside the plant tissues,
such as bollworms and root-borers, which were hard to
reach with sprays. But, unlike chemical sprays, it would

not affect harmless insect species with no desire to eat the crop plant. It proved to be a triumphantly successful innovation. Almost every cotton garment you buy today is a product of such a genetically engineered plant: over 90 per cent of the cotton grown in the world is insect-resistant. In India and Pakistan, the technology was rapidly adopted by farmers while still illegal, as its benefits became dramatically obvious elsewhere in the world. It was then legalized, and today almost all cotton grown in the two countries is Bt.

About one-third of the maize (corn) grown in the world is now insect-resistant because of introduced Bt genes as well. In America, where 79 per cent of the corn is now Bt, the cumulative benefit to farm income of this technology over twenty years comes to more than $25bn. Bizarrely, the organic-farming sector refused to approve the new plants even though they used the same molecules as their own sprays, because of an objection to biotechnology in principle.

Because a Bt crop is protected without much if any spraying, there has been a noticeable increase in wildlife on farms that adopt the Bt technology, as well as a reduction in accidental poisonings of farmers themselves by sprays. In some Chinese studies, a doubling of natural insect predators such as ladybirds, lacewings and spiders was recorded in Bt cotton fields, meaning better control of all pests by natural predators. Research at the University of Maryland has now found that Bt crops create a 'halo effect', in which surrounding crops and fields, not growing Bt crops, also have reduced pest problems. In the twenty years since the introduction of Bt crops, the populations of two common pests, the European corn borer and the corn earworm, both of which attack other plants as well as maize, have fallen so much in three American states that even organic and non-genetically-modified farmers are able to use less spray than before: 85 per cent less on peppers. Overall, a comprehensive study of the effect of Bt technology concluded that after a billion acres had been planted, there were zero unintended consequences, and large benefits for non-target insects.

This technology is proving especially useful in developing countries. Africa is currently facing an intense crisis because of the arrival on the continent in 2016 of a pest from the Americas – the fall armyworm – which is now devastating maize crops across the continent. The pest is no longer a problem in Brazil because of the use of Bt maize there, but African countries, under pressure from well-funded ideological opponents of genetically modified crops, have been slow to allow Bt maize to be grown.

These opponents had been especially successful in Europe, discovering in the late 1990s that spreading scare stories about genetically modified crops among easily spooked consumers was a lucrative way of raising funds. To Van Montagu's dismay, Europe rejected the technology almost entirely, by erecting high and costly regulatory barriers to its deployment, which amounted to a de facto ban (see chapter 11).

All pest control eventually runs into the evolution of resistance in the pest, though this has been less of a problem with Bt crops than with pesticides. However, the latest generation of Bt crops includes sophisticated extra features that ensure that insects will be much slower to develop resistance to the Bt protein. So the innovation path that led from the discovery of a bacterial disease in silkworms more than a century ago has led to dramatically reduced crop loss, pesticide use and environmental damage. Most crops are now also herbicide-tolerant so that they can be combined with effective weed control without the soil-damaging practice of ploughing. Some are also being engineered to be resistant to fungal disease or drought. Others are being engineered to fix their own nitrogen with the help of bacteria, greatly improving yields. Yet others are being engineered to remove a metabolic penalty found in all 'C3' plants (which include wheat, rice, soybeans and potatoes, but not maize), whereby oxygen diverts the photosynthetic machinery into producing a wasteful product. The first such modified tobacco plants had a 40 per cent greater yield and flowered a week earlier in field trials published in 2019.

Gene editing gets crisper

Highly useful scientific discoveries are almost always – ridiculously often – accompanied by frenzied disputes about who deserves the credit. In no case is this more true than in the story of CRISPR, a genetic technique that the world awoke to in 2012, and which promises wonderful results in agriculture as well as medicine. The dispute is sharpened in this case by the fact that it pits two great American universities, on opposite coasts, against each other. There is Berkeley in California, where Jennifer Doudna worked, while collaborating with Emmanuelle Charpentier, a French professor who had recently moved from Vienna to Umeå in Sweden, and Charpentier's graduate student Martin Jinek. And there is MIT in Massachusetts, where Feng Zhang and his colleagues Le Cong and Fei Ann Ran worked. Both groups had crucial breakthroughs around the same time. Initially more prizes went to Doudna's group, but a fiercely fought patent battle was eventually won in the courts by Zhang's group.

Yet arguably neither of these huge American universities with their big budgets and luxurious laboratories deserve as much credit as they seek. That should go to a couple of obscure microbiologists working on practical but unfashionable questions about bacteria, one in a university laboratory tackling a problem of interest to the salt industry, the other in an industrial food-manufacturing company. The road from the discovery of a biochemical curiosity to the invention of a technology, as ever, is long and winding. And in this case it goes not from academia to industry but at least partly in the opposite direction.

Near the town of Alicante, in Spain, is a large, pink lake, dotted with even pinker flamingos. Known as Torrevieja, this 1,400-hectare lake lies below sea level and has been used for three centuries to produce salt. In June, sea water is allowed to flow into the lake. Over the summer as the water evaporates, salt crystallizes on the floor of the lake and is scooped up by special machines to be cleaned and sold – 700,000 tonnes of it a year. The pink colour comes from salt-loving microbes, of two kinds, bacteria and archea, which are eaten by pink shrimps, which are eaten in turn by pink flamingos.

Not surprisingly, the local university's microbiology department has made use of this resource to study salt-loving pink microbes. One archeal microbe, called *Haloferax mediterranei*, was first described in Alicante. Perhaps, being such a salt-loving species, it could be used for biotechnology in especially salty places. Francisco Mojica, who

was born nearby, earned a doctorate here in 1993 studying the genes of this creature. And he noticed something
rather odd. Hidden in part of its genome was a distinct-
ive sequence of the same thirty letters, repeated over and
over again, each repeat being separated by a sequence of
35–39 letters that was different in every case. The repeat
sequence was often a palindrome – it spelled the same text
backwards and forwards. Mojica looked in another, related
salt-loving microbe and found roughly the same pattern,
though with a different sequence. He then found it again
in twenty different microbes, both bacterial and archeal.
A Japanese researcher had spotted the same pattern in a
bacterium in the 1980s but had not followed it up.

Mojica spent the next ten years trying to understand
why this pattern was there. Most of his hypotheses proved
wrong. A Dutch scientist, Ruud Jansen, noticed that there
were always certain genes near the strange text, known as
Cas genes. Jansen coined the name for the pattern: 'clustered
regularly interspaced palindromic repeats', or CRISPR for
short.

Then, one day in 2003, Mojica had a lucky break. He
took one of the non-repetitive 'spacer' sequences, between
the palindromes, from a gut bacterium, and put it into a
database of gene sequences to see what it matched. Eureka.
The answer came back: it matched the gene of a virus,
specifically a bacteriophage virus, known as 'phage' for
short. These tiny particles, sometimes shaped like minuscule lunar landers from an Apollo mission, are viruses that
inject their DNA into bacteria, hijack their cellular machinery and make more phages. Mojica looked at more spacer
sequences and found that many of them came from viruses
that infect bacteria. He surmised that he was looking at a

microbe's own immune system, in which genes of viral diseases were kept on file by the microbe for recognition and destruction. The Cas genes do the work.

It took Mojica more than a year to get his results published, so sniffy were the prestigious journals at the idea of a significant discovery coming from a scientific nobody in a backwater like Alicante. Across the Pyrenees in France, an industrial microbiologist was already taking the next step. Philippe Horvath worked for Rhodia foods, which soon became part of Danisco and later part of DuPont. Yogurt and cheese are fermented milk: they depend on bacteria to eat the milk and convert it into bacterial bodies, which is what we eat. The microscopic domesticated milch-cow of the dairy industry is a harmless creature called *Streptococcus thermophilus*. The average person eats about a thousand billion billion *S. thermophilus* a year. Big companies that make yogurt therefore spend a lot of money on bacteriology the better to understand their domesticated flocks of microbes. They are especially interested in what happens when the bacteria get sick. Just as a dairy farmer wants to protect his cows against mastitis, so a yogurt maker needs his streptococci not to get infected with 'phages'. Horvath and his Danisco colleague Rodolphe Barrangou knew that some cultures of bacteria prove to be more resistant to phage epidemics than others: understanding why this was the case might help the industry.

After hearing about CRISPR at a conference, Horvath had a hunch that it might supply the answer. He soon showed that the bacteria with the most spacers were often the most likely to be the resistant strains, and the ones with spacers derived from a particular phage DNA were resistant to that phage. This proved Mojica right. CRISPR's job

– with the help of Cas – is to recognize a particular sequence and cut it, thus emasculating the virus.

The next step, or leap of logic, was to think 'maybe we human beings can borrow CRISPR for our own purposes'. Replace the spacers with a gene we want to excise, perhaps combine it with a new sequence we want to insert, and adapt the microbial system as a genetic-engineering tool of uncanny precision. Instead of waiting for nature to throw up better genes, as we did in the 1920s, or mutating genes at random using gamma rays, as we did in the 1960s, or throwing in specific new genes in the hope in some cases they would land somewhere useful, as we did in the 1990s, now we could literally edit the genome of a plant or animal, changing one letter here, or one sentence there, using the CRISPR-Cas9 system. Gene editing was born.

In 2017 scientists at the Roslin Institute, near Edinburgh, announced that they had gene-edited pigs to protect them against a virus called Porcine Reproductive and Respiratory Syndrome virus (PRRS). They used CRISPR to cut out a short section from the gene that made the protein that gave the virus access to pig cells, thus denying the virus entry, and they did this without altering the function of the protein so the animal grew up to be normal in every way, but immune to the disease. In 2018 scientists from the University of Minnesota and Calyxt, a genetic company, used a different gene-editing technique called TALEN to make a wheat resistant to powdery mildew, so it needs less fungicide. That same year Argentinian scientists used CRISPR to snip out part of the polyphenol oxidase gene in a potato, making the potato incapable of turning brown when cut. As of mid-2019, there are over 500 gene-editing projects underway in China, nearly 400 in America and almost 100 in Japan. (Most of these relate to agriculture, though of course gene editing will be applied in medicine as well.)

And Europe? Most of the world quickly agreed that gene-edited plants should not be subject to the same immensely expensive and delaying regulation as GMO crops, but be treated like conventionally bred varieties. All across Europe scientists hoped and prayed that the same conclusion would be reached by the authorities there. The European Commission waited two years for the European Court of Justice to opine. The court's advocate-general argued for liberalization, but in July 2018 the court, under political pressure, rejected his advice and ruled that gene-edited organisms must be treated to the same regulation as GMOs, not the much simpler rules applied to mutagenesis crops, those treated with gamma rays or chemical mutagens in a far more risky process.

In 2019 three French scientists reviewed the patenting of CRISPR products and found that Europe was already being dramatically left behind. Whereas America had taken out 872 patent families and China 858, the EU had only taken out 194 and the gap was growing. They concluded: 'It would be a delusion not to consider the GMO bans in Europe as having had a strong negative impact on the future of biotechnology on the continent.'

Gene editing is changing fast. Already, base editing, or prime editing – in which DNA bases are chemically replaced without snipping the DNA strand – has begun to appear, with far greater accuracy than gene editing. There is no doubt that extraordinary improvements in the yield, nutritional quality and environmental impact of food crops is going to be possible in the future.

Land sparing versus land sharing

The immense improvement in the yield of farming during the twentieth century, as a result of innovations in mechanization, fertilizer, new varieties, pesticides and genetic engineering, has banished famine from the face of the planet almost entirely, and drastically reduced malnutrition, even while the human population has continued to expand. Few predicted this, yet many are concerned that this improvement has come at the expense of nature. In fact the evidence is strong that the opposite is the case. Innovation in food production has spared land and forest from the plough, the cow and the axe on a grand scale by increasing the productivity of the land we do farm. It turns out that this 'land sparing' has been much better for biodiversity than land sharing would have been – by which is meant growing crops at low yields in the hope that abundant wildlife lives in fields alongside crops.

Between 1960 and 2010, the acreage of land needed to produce a given quantity of food has declined by about 65 per cent. Had this not happened, pretty well every acre of forest, wetland and nature reserve in the world would have been cultivated or grazed, and the Amazon rain forest would have been far more severely destroyed. As it is, the acreage of wild land and nature reserves is steadily increasing, while forest cover has stopped declining and in many places is now increasing, so that overall there has been a 7 per cent increase in tree cover since 1982. By the middle of the current century, the world will be feeding nine billion people from a smaller area of land than it fed three billion from in 1950. Moreover, recent studies have concluded that – for a given yield of food – intensive agriculture not only uses less land but produces fewer pollutants, causes less soil loss and consumes less water than organic or extensive systems.

Now imagine that innovation continues to improve the yields of farms, by tweaking the efficiency of photosynthesis, inserting nitrogen-fixing bacteria into plant cells, further reducing losses to insects, fungi and weeds and diverting still more of each plant's energy into valuable food (all of which are happening), so that the average yields of crops like rice, wheat, maize, soy and potatoes are 50 per cent higher in 2050 than they are now. This is definitely plausible, maybe even probable. That would mean we could cultivate much less land, enlarging national parks and nature reserves, returning land to forest and wilderness, managing more land for flowers, birds and butterflies. We could enhance the ecology of the planet even as we feed ourselves.

5

Low-technology innovation

When zero is added to a number or subtracted
from a number, the number remains unchanged;
and a number multiplied by zero becomes zero.

BRAHMAGUPTA, the year 628

When numbers were new

'These are the nine figures of the Indians: 9 8 7 6 5 4 3 2 1.
With these nine figures, and with this sign 0 which in
Arabic is called zephirum, any number can be written, as
will be demonstrated.' Thus did an Italian merchant around
the year 1202 (or MCCII) introduce Europe to modern
numerals, modern arithmetic and crucially the use of zero.
Leonardo of Pisa, today known by the nickname Fibonacci,
had travelled as a child from Pisa to Bugia, a port on the
coast of North Africa, where his father was the diplomatic
representative of the Pisan traders who imported wool,
cloth, timber and iron into North Africa, while exporting
silk, spices, beeswax and leather to Genoa.

Fibonacci learned arithmetic in Bugia in the Arab style, and probably in the Arabic language, and he quickly realized that the Arabic notation, borrowed from the Indians, was far more practical and versatile than Roman numerals. 'There from a marvellous instruction in the art of the nine Indian figures, the introduction and knowledge of the art pleased me so much above all else, and I learned from them, whoever was learned in it, from nearby Egypt, Syria, Greece, Sicily, and Provence, and their various methods, to which locations of business I travelled considerably,' boasted Fibonacci later.

There are two features of Indian numbering that are astonishingly helpful. One is the idea that the position of a number in a sequence indicates its size. So 90 is ten times bigger than 9, whereas X means ten in Roman numerals wherever it appears in a number. The other feature is that this positional system only works in decimal systems if one of the ten numerals stands for nothing. The language of mathematics, writes Robert Kaplan, 'comes into its own when zero entered it as the sign for an operation: the operation of changing a digit's value by shifting its place'.

But a symbol for nothing is bafflingly counterintuitive if you stop to think about it. Nothing of what? As Alfred North Whitehead put it: 'the point about zero is that we do not need to use it in our daily operations. No-one goes out to buy zero fish.' (Though I sometimes go out fishing and catch zero fish.) Zero turns numbers from adjectives into nouns, and becomes a number in its own right. This was an innovation of far-reaching consequence, for sure, but it involved no technology.

Considering how indispensable Indian numbers are to modern life, and how impossible it would be to live without them, this innovation was extraordinarily important, and it is bizarre how late it enters the story of Western civilization. The whole of the classical world and early-medieval Christendom got by with a system of counting that made multiplication virtually impossible, algebra unfathomable and accounting primitive. Fibonacci's role in this revolution was forgotten until late in the eighteenth century, when a scholar by the name of Pietro Cossali, studying the work of a great fifteenth-century mathematician, Luca Pacioli (a close friend of Leonardo da Vinci), noticed that Pacioli in passing mentioned that 'we follow for the most part Leonardo Pisano'. Cossali sought out this earlier Leonardo's manuscripts and realized that almost all of the mathematical treatises of the intervening centuries were derived more or less directly from his hefty tome, the *Liber abbaci*. The name 'Fibonacci' was coined in the nineteenth century as a contraction of the phrase 'filius Bonacci' or son of a good bloke, which appeared on the title page of his book. Since *Liber abbaci* appeared two centuries before the printing revolution, and therefore relied on being transcribed, its success as a manuscript had been lost in the mist of time.

Fibonacci's work was one of the most influential compositions in all European history, garnering him an audience with the Holy Roman Emperor, the intellectually curious but cruel Frederick II, as well as being copied and disseminated all over Europe till Indian numerals had almost entirely displaced the Roman species. The irony is that Indian numbers were not wholly unknown on the northern shores of the Mediterranean, but they were a scholarly speciality, especially in Spain, where Christian monks had borrowed them from the Arabs but only to study

mathematics. The works of Al Khwarizmi, the great ex-
positor of algebra, had been translated into Latin but for
scholars, not merchants.

What Fibonacci did was show merchants how to use
this arithmetic in everyday commercial transactions. His
book was filled with practical questions, each redolent of
this world of Mediterranean trade dominated by the Italian
city states and their trading partners in the Near East and
Maghreb. Such as: 'If one hundredweight of linen or some
other merchandise is sold near Syria or Alexandria for 4
Saracen bezants, and you will wish to know how much 37
rolls are worth . . .' Note that this was after the first three
crusades, and around the time of the fourth, so plenty of
Christian chiefs and priests were living, ruling and fighting
in the Near East, but it was a merchant who got the message across. It is notable that this innovation, like so many,
comes to us through commerce.

Fibonacci, however talented an innovator, was not the
inventor, but the messenger. (He did invent plenty of mathematics, including the famous Fibonacci series and the
golden ratio derived from it, found throughout growing
organisms in nature – such as the shell of a snail or the
seeds of a sunflower – but he did not invent Indian numbers or zero.) His sources were Arab, and the greatest of
them was Al Khwarizmi, the mathematician whose name
survives into English in the word 'algorithm'. Fibonacci
read his work in Latin translation when back in Italy as
well as probably in Arabic. Yet Al Khwarizmi, too, was not
the inventor of much of this, but the compiler and popular-
izer, as the title of his most important work shows: 'On the
Calculation with Hindu Numerals', published around the
year 820. He had played a role in the Muslim world not

unlike Fibonacci in the Christian world: he aimed his book at merchants and he was explaining an innovation that his civilization had borrowed from another.

Tracing the trail two centuries further back, to the year 628, we find Brahmagupta, an astronomer living in a kingdom of western India called Gurjaradesa, known for its scholarship. He published a book called the *Brahmasphuta-siddhanta*, or the 'opening of the universe'. Though mostly about astronomy, it had chapters on mathematics and is the first known work to treat zero as an actual number, rather than as a symbol for nothing as the Babylonians had done. In simple and easily understood statements, Brahmagupta set out the significance of zero and considered negative numbers for the first time, driving the point home in homely terms: 'A debt minus zero is a debt. A fortune minus zero is a fortune. Zero minus zero is a zero. A debt subtracted from zero is a fortune. A fortune subtracted from zero is a debt. The product of zero multiplied by a debt or fortune is zero.' After that the trail goes cold. The oldest written use of zero as a placeholder – when it was a dot – is found in the Bakhshali manuscript of the fourth or fifth century AD, which was discovered in what is now Pakistan in 1881. Something like it had been used in ancient Sumer and Babylon, whence it might or might not have travelled east with the Greeks who followed Alexander to India. But there is no evidence before Brahmagupta that zero was being used in its present, numerical form and thus transforming arithmetic.

But hang on. In the best tradition of parallel innovation, there is evidence that the Mayans invented zero around the same time as Brahmagupta, perhaps earlier. In their 20-based counting system used for the Mayan long calendar there was a glyph that stood as a spacer, some-what like the Hindu 0. It proved to be a dead end. The Mayan civilization collapsed and took its best arithmetical idea with it. Could the same have happened in the Old World? Fibonacci was a contemporary of Richard the Lionheart, Saladin and Genghis Khan, all bloodthirsty war-riors. Warfare, religious fanaticism and tyranny were on the march. Two great capitals of learning had recently turned their back on freedom of thought in favour of mysticism: Baghdad under Al-Ghazali, and Paris under St Bernard of Clairvaux. India, too, was a battleground between Islamic and increasingly fundamentalist Hindu dynasties. China was crushed by Mongol armies. Perhaps it was just as well that Fibonacci took zero across the sea to Pisa and the other city states of northern Italy, where commerce thrived and people cared more for practical enterprise, for buying low and selling high, rather than glory or God.

Fibonacci's innovation co-existed with other ways of counting and accounting for centuries: counting boards, tally sticks, the abacus. Even on paper it sat alongside Roman numerals. In the fourteenth century, ledgers would sometimes have columns of Indian numbers and paragraphs of Roman ones intermingled or taking turns. Yet gradually numbers won, especially in the preparation of merchants' accounts: commerce led the way. By the time Luca Pacioli wrote his great treatise on double-entry book-keeping in 1494, making clear just how vital Fibonacci's innovation was for mathematicians as well as accountants, Roman numerals were used mainly for dates and monuments. As they still are today: I have seen people write 7.ii.19 for a date atop a letter.

The water trap

I walk a lot in London, and a few months ago I set myself a goal: somewhere in the vast city, while walking down a street, to catch the smell of sewage. I have yet to achieve this goal. Close to ten million defecations occur in London every day, presumably, since for most people it is a daily occurrence. I hazard that I am rarely more than 100 feet from somebody actively at work on this task. According to the Parliamentary Office of Science and Technology, the volume of sewage produced in London is more than a billion litres every day: 400 billion litres a year, or enough to fill ten million standard swimming pools.

Yet you never smell it. Why not? This is a new phenomenon, an innovation. In past eras, cities smelled richly of sewage all the time, and you would be hard pressed to walk down a street without seeing it or stepping in it, let alone smelling it. Today, sewage is still there, all around us, yet kept so entirely separate from us that we never even smell it, let alone see it. It is taken away, treated and disappears, almost wholly unseen. When you think about it, this is quite an achievement, one of the finest of our civilization.

Lots of innovations contribute to this, most of them simple and low tech, such as sewers themselves. Perhaps the neatest innovation is the S-bend or U-bend in the pipe beneath every toilet, which traps water so as to prevent any smell coming back up the pipe. It's gorgeously simple and exquisitely clever. It transformed the flush toilet into a strong competitor against the chamberpot. Flush toilets were tried many times before, beginning with a device invented in 1596 by Sir John Harington, a godson of Queen Elizabeth I, who had one installed in Richmond Palace. Harington even wrote a book about it with the punning title, *The Metamorphosis of Ajax*, 'jakes' being a contemporary term for toilet. The queen had the book hung from the wall in the privy presumably as loo reading. But it did not catch on. Flush toilets were expensive and unreliable and they had the huge disadvantage that they took away the sewage but not its smell. Carrying a chamberpot out of the building worked much better in that regard.

The S-bend is one of those things that could have been invented at almost any time and by almost anybody. It ought to be the classic case of a journeyman plumber doing something that had evaded brilliant thinkers. Yet, surprisingly, it is the product of a fine mathematical mind at the height of the Enlightenment. His name was Alexander Cumming and he was mainly a maker of clocks and organs, though he also wrote disquisitions on carriage wheels and dabbled in pure mathematics.

We know nothing of his origins beyond the fact that he was born in Edinburgh and came to London to win the patronage of King George III, for whom he made an ingenious pendulum-driven barometer clock that recorded the air pressure on a paper chart. His chronometers were so good that the Arctic explorer Constantine Phipps named a small island after him north of Spitsbergen.

Apart from that, there is not a lot to say about Mr Cumming. He was granted a patent on 'a water closet upon a new construction'. It included many of the features we know today, most critically the S-trap. It flushed from an overhead cistern, and a little water remained in the double bend of the pipe to act as a barrier to smell. However, Cumming included a feature that was quite unnecessary and proved troublesome: a sliding valve across the base of the bowl, above the S-trap, that had to be opened and closed by a lever. This leaked. It also jammed, especially in frosty weather (and most toilets were out of doors in privies in those days), or when it had rusted or become encrusted with scale. So Cumming, like Harington, saw his invention only slowly adopted.

Three years later in 1778, the water closet was transformed by another innovator, Joseph Bramah. The son of a Yorkshire farmer, born in 1749, Bramah has a string of inventions to his name in many different fields. His most important was hydraulics, so crucial to much machinery today, though in fact it was his even more talented employee, Henry Maudslay, who contributed the crucial ideas. His most famous is the Bramah lock, also built by Maudslay. It was virtually impossible to pick, as demonstrated by Bramah's firm offering a prize of 200 guineas to the first person to do so. It remained unclaimed for nearly half a century, till 1851, long after Bramah's death, and then only when an American lock-picking impresario named Alfred Hobbs spent more than a month achieving the feat with a clutch of specially made instruments. By then Bramah's firm had a new version of the lock.

After a leg injury as a teenager left him permanently lame and unfit for farm work, Bramah discovered a talent with woodworking, served his apprenticeship as a joiner and moved to London to work as a cabinet maker. He was employed by a Mr Allen, who was probably used by Cumming to make a cabinet to hold his water closet. Allen improved the water closet by causing the water to spiral around the bowl when flushed. Around this time Bramah had another accident, and while laid up turned his mind to improving the water closet further. He patented his design in 1778, with a hinged flap instead of a sliding valve and a series of other tweaks to the design. What is more, he brought his own exquisitely high standards of craftsmanship to the product, and it began to sell. Bramah set up in business and was soon installing six water closets a week for the wealthy, at over £10 a time. The success of the product was proved by the fact that others soon copied it, and Bramah took several to court. One case in 1789 set legal precedent. The defendant, a Mr Hardcastle, argued that Bramah's patent was too vaguely worded, included features that were not novel and crucially had been 'published' before. The argument on this latter point was that Bramah had built three water closets to his design and tested them before applying for the patent. The judge ruled in favour of Bramah, stating – perhaps from practical experience – that the design worked better than any previous one.

The indoor water closet really only took off as a must-have item towards the end of the nineteenth century. The building of a vast new sewer system in London meant that at last water closets had somewhere to send waste to, even from the humblest house. The antipathy of many people to having WCs indoors began to change. Thomas Crapper, a Yorkshire plumber who set up shop in London in the 1860s, was an entrepreneur who capitalized on this new demand. He invented little, but improved the water trap by making it more of a U-bend rather than an S-bend, rendering it less likely to block. He improved the siphon system from the cistern and the ballcock mechanism (a British peculiarity) to prevent the cistern overflowing. But his real achievement was to make water closets reliable, simple and affordable, and lend his very name to them – even though strangely the verb 'to crap' was much older.

Crinkly tin conquers the Empire

Disliked for its ugliness, overlooked for its ordinariness, and so old it is hard to think of it as an innovation: corrugated iron is an unlikely hero. Yet it was a novelty once – invented in 1829 – and has arguably been a greater benefit to human beings than many a more glamorous thing. It has sheltered countless millions of people from the rain and the wind, and has done so more cheaply and effectively than far more celebrated architecture. It has kept the poor alive in shanty towns, favelas and slums. In the form of Anderson shelters, it has saved lives in bombing raids. In California, Australia and South Africa it was indispensable for gold miners erecting instant towns. In Australia it proved to be popular with both settlers and natives, who called it 'the white man's bark'. At one point it was so fashionable that architects built churches out of it. Prince Albert added a ballroom to Balmoral in corrugated iron.

As innovations go, the story of corrugated iron is comparatively simple. It appears to have been invented by one person, unchallenged by rivals. He was a trained engineer, not an obscure genius or a brilliant scientist. His patent went unchallenged and when it expired the product quickly grew and grew as an export industry. It was improved at various points in history, mainly to make it more resistant to corrosion, but it remains essentially the same design today as it was at the start.

The inventor was Henry Robinson Palmer. His other ideas were both far ahead of their time: the monorail and containerization. Born in 1795 in east London, the son of a parson, he apprenticed as an engineer, worked for ten years for the great civil engineer Thomas Telford and was a founder of the Institution of Civil Engineers. In 1826 he was appointed to oversee the extension to a dock in east London. Having finished the excavation and construction of the locks he turned his attention to the buildings. He seems to have hit upon the idea of using an iron sheet for the roof of an open shed, but to make the sheet stronger, he passed the wrought iron through rollers to give it a sinusoidal wave. On 28 April 1829 he patented 'the use or application of fluted, indented or corrugated metallic sheets or plates to the roofs and other parts of buildings.' This immensely strengthened the iron sheet, made it more rigid and capable of spanning a wide gap without extra support while supporting a load such as snow. Crinkly tin was born.

At the dock the corrugation was done on site and the first building erected with a curved, self-supporting cast-iron roof. George Hebert, editor of the *Arts and Sciences*, visited shortly afterwards and was much taken with 'Mr Palmer's newly invented roofing'. The 'grooving, or as we might say, arching and counter-arching, confers great strength', Hebert accurately reported, so that a sheet of metal just one-tenth of an inch thick could provide a sturdy roof spanning eighteen feet: 'It is, we should think, the lightest and strongest roof (for its weight), that has been constructed by man, since the days of Adam.'

Corrugated iron has evolved continuously since then with scores of patents on improvements. For instance, within ten years the process of galvanization, invented by Stanislas Sorel in France, protected iron from rusting with a thin layer of zinc and gave corrugated iron a much longer lifespan. Later in the century, steel replaced wrought iron as the main ingredient. But the basic design barely changed.

Palmer sold the patent to his assistant, Richard Walker, who, together with his sons, was to dominate the industry for decades. He grew wealthy before the patent expired in 1843. Only after that did the market rapidly expand as the price came down. Intellectual property therefore merely served to delay the innovation, as usual.

By 1837 Walker was advertising corrugated iron for use in Australia, a continent that would come to embrace the material more than any other. 'Australia is, beyond doubt, the spiritual home of corrugated iron,' wrote Adam Mornement and Simon Holloway in their 2007 history of the material. Its resistance to termites and fire, its light weight and its prefabricated nature in a country with scarce labour – all these recommended corrugated iron to the colonists of the Australian continent. The goldrush of the 1850s in Victoria resulted in growing demand for quick-fix, new building materials and soon entire towns of corrugated iron were springing up in the gold fields. In 1853 Samuel Hemming shipped a complete church from London to Melbourne for £1,000, from where it was transported to Gisborne by bullock cart and erected for a further £500.

By 1885 Australia was the largest market for the stuff in the world, and in the 1970s it was an Australian firm, BHP, that patented Zincalume steel, a corrugated material made of steel, but coated in 55 per cent aluminium, 43.5 per cent zinc and 1.5 per cent silicon. This is more resistant to corrosion than normal zinc-coated steel. Recently, corrugated iron's place in Australian history has made it a trendy material for architects and artists: the opening of the Sydney Olympics included a specially composed 'Tin Symphony' in its honour, while the artist Rosalie Gascoigne used the material in her sculptures.

From Australia, the habit of building in corrugated iron spread to Africa, where the gold mining boom of South Africa in the late 1800s depended heavily on corrugated iron, manufactured in Australia, shipped to Durban and carried inland by teams of porters to make anything and everything: roofs, walls, water tanks, whole buildings. In the Boer War the British built blockhouses of double skins of curved corrugated iron, the space filled with shingle, to defend railways. From the trenches of the First World War to the whaling stations of South Georgia, corrugated iron was a vital part of twentieth-century construction. The Nissen hut, a semi-cylindrical shelter of corrugated iron on a steel frame, invented by Norman Nissen, an American engineer, proved a cheap, safe and quick building in both world wars.

In the slums of today's expanding megacities, where property rights are uncertain, corrugated iron is not only affordable and available, but buildings made of it can be easily dismantled and moved. It is one of the first things shipped into earthquake zones to provide shelter in short order. It has probably also saved a lot of forests, since it requires so much less timber support than many other building materials. It may never be loved or admired, and the drumming of rain on roofs made of it may not be the sweetest of sounds, but it was a simple innovation that certainly changed the world.

The container that changed trade

The *Warrior* was a normal cargo ship contracted by the US military to carry a typical 5,000-ton cargo on an unremarkable voyage from Brooklyn to Bremerhaven in Germany in 1954. The cargo consisted of 194,582 items – cases, cartons, bags, boxes, bundles, packages, pieces, drums, barrels, crates, vehicles and more. They arrived in Brooklyn in 1,156 shipments from 151 American cities. Loading took six days including one lost to a strike. The voyage took almost eleven days. Unloading took four days. Port costs accounted for 37 per cent of the total shipping cost of $237,577, whereas the sea voyage itself cost just 11 per cent. We know all this because of a government-sponsored study of this one cargo, cited by Marc Levinson in his book about the invention of container shipping, *The Box*. The study's conclusion was that in tackling the high costs of ports, 'perhaps the remedy lies in discovering ways of packaging, moving and stowing cargo in such a manner that breakbulk is avoided.' Within a few years containerization was an innovation that transformed the world. It was a momentous innovation but it involved no new science, no high technology and not much new low technology, just a lot of organization.

In the mid-1950s shipping goods by sea was almost as expensive, slow and inefficient as it had been for centuries. Despite faster engines and bigger ships, the ports were costly bottlenecks. More than half the cost of exporting or importing consisted of port costs (the *Warrior* voyage was unusually good value in this respect, because of low labour costs in post-war Germany). The dockers or longshoremen who handled the work earned relatively good wages for manual workers, but the job was labour-intensive, dangerous, uncertain, irregular in hours and exhausting. Cargoes were deposited on the quayside, sorted, stored in warehouses, piled on pallets, slung by cranes aboard ship, unloaded from pallets and stowed largely by hand into holds that were usually curved and variable in shape, making the securing of a cargo as much an art as a science. Forklift trucks and cranes helped, but a great deal of elbow grease did most of the work. The whole process was repeated on arrival at the other end, with customs inspections added in. International trade as a percentage of the economy in the United States had actually been shrinking since the 1920s largely because of the costs incurred in ports. Union closed shops had recently done away with the bribery and violence that accompanied the scramble for irregular work at docks, but at a cost of higher charges. The quantity of cargo handled by a single man in a year fell during the 1950s in the ports of Los Angeles, New York and London, even as wages rose.

The idea of standard containers, boxes of uniform size and shape pre-loaded at factories with goods and lifted on and off ships without being opened, was not new. Railways had been experimenting with standardized containers for decades, and trucks too. An American firm called Seatrain Lines had started using specially designed ships to carry railway boxcars in 1929. But the results had so far been disappointing. The containers were either too big to fill quickly, so they sat around at factories, or too small to be much help, their own weight adding to the cost of the cargo. They wasted space by not fitting neatly into holds or being half empty. 'Cargo containers have been more of a hindrance than a help,' a leading shipping executive concluded in 1955 just in time to be proved dramatically wrong.

Then came Malcom (sic) McLean. Born in 1913 in a landlocked town of largely Scottish-descended folk, Maxton in North Carolina, McLean was one of those ambitious, risk-ready entrepreneurs who make getting rich look simple. Working at a fuel station, he realized there was good money to be made transporting fuel, so in 1934 he borrowed an old tanker and started trucking. Within a year he owned two trucks and employed nine drivers with their own trucks. By 1945 his business had 162 trucks and earned $2.2m. McLean knew how to get round the fussy interstate commerce regulations, and his mostly self-employed drivers were less prone to strike than those of his competitors. They earned bonuses for not having accidents, which kept repair costs down. To save money he switched to diesel early, and pioneered conveyors to move cargo between trucks. By 1954 he had over 600 trucks, financed by a lot of debt.

By then he had had an idea. Coastal shipping was in decline, not having recovered from the war, whereas the roads were increasingly congested. Why not drive his trailers on to ships and have rigs pick them up at ports nearer the destination? With a high appetite for risk, he sold the truck business and bought a big shipping business instead, with borrowed money – effectively inventing the leveraged buyout. But then he had a better idea. Instead of putting whole trailers on ships, why not lift the bodies of the trailers off the wheels and stack them on the ships instead? He tested the plan on paper with a shipment of beer from New York to Miami and found it could cut the cost by 94 per cent compared with break-bulk cargo.

The legend that grew up around this story is that McLean, like Archimedes and Newton, had a sudden moment of inspiration while waiting for a truck to be unloaded at a port back in the 1930s. Like all such stories, it is false, though hard to kill. As Levinson recalls, having studied the history:

To my consternation, though, I quickly learned that many people quite fancy the tale of McLean's dockside epiphany. The idea of a single moment of inspiration, of the apple landing on young Isaac Newton's head, stirs the soul, even if it turns out to be apocryphal. In contrast, the idea that innovation occurs in fits and starts, with one person adapting a concept already in use and another figuring out how to make a profit from it, has little appeal.

Why do such heroic myths persist? Perhaps the truth is that people like to think they too could become heroes with a single leap of imagination. Such magical thinking is deeply misleading as to the character of most actual innovators. The facts are less remarkable, but more daunting, and McLean is a case in point.

McLean bought an oil tanker, the SS *Ideal X*, and converted her to carry containers on a specially designed deck; he bought two big cranes and converted them to lift containers; and he commissioned the construction of a fleet of 33-foot containers. He then spent two years persuading the authorities in the Interstate Commerce Commission and the Coastguard that the ship was safe, while fighting off spoiling actions in the courts by railways and truckers. On 26 April 1956 the *Ideal X* set sail from New Jersey to Texas with fifty-eight containers on board. It took seven minutes to lift each container on board, and just eight hours to load the ship. By the time the voyage was over, McLean reckoned it had cost less than 16 cents a ton, compared with $5.83 a ton for normal cargo rates.

Such enormous cost saving spoke for itself, or so one would think. But McLean's battle was only beginning.

The engineering bit went smoothly at first. He used a dock strike in 1956 to redesign six larger ships to hold 226 containers each at his headquarters in Mobile, Alabama. By trial and error, his engineer, Keith Tantlinger, worked out just how much tolerance to allow in the metal cells in the hold where the containers sat: just over an inch in length and just under an inch in width, enough to make loading simple, not enough for the container to shift in a storm. (Tantlinger used modelling clay stuffed into the space around the containers to prove that they did not shift on the first voyage.) Systematically, Tantlinger redesigned everything from the truck chassis to the containers themselves and the twist-lock that held them together on board to make them quicker to load and unload. The new gantry cranes on board could even load and unload a ship at the same time. The first of these ships, the SS *Gateway City*, built in Mobile in 1957, could be loaded and unloaded in eight hours, the same time as the *Ideal X*, though she carried five times as many containers.

The chief obstacles McLean encountered were human ones. In 1958 he sent two of his new ships from Newark to Puerto Rico, where the longshoremen's union refused to unload them. They sat idle for four months till McLean caved in and agreed to unnecessarily large crews of men unloading each ship. The delay cost him all the previous year's profits. Another strike in 1959 caused further losses and brought McLean's business to the brink of bankruptcy. Other shipping companies baulked at the hefty investment needed to get into container shipping, especially with an uncooperative labour force, so ports were reluctant to change. The container revolution seemed like a failure.

McLean responded by hiring hungry, young, entrepreneurial people from his former trucking business for the renamed Sea-Land business to crack the problems. He borrowed more money and built even bigger ships. He started shipping from the East Coast to California, through the Panama Canal. He had a stroke of luck when his main competitor on the Puerto Rico route went bust after its buyer took on too much debt. By 1965 Sea-Land had fifteen ships and 13,533 containers. After a long internal battle the unions eventually came round to mechanization, as it brought more business to ports and better working conditions. On the West Coast unions even argued that employers were dragging their feet in automating work.

The key problem now was standardization. The United States government and then the International Standards Organization wrestled for years with what would be the best size and shape for a 'standard container'. By 1965, though, two-thirds of containers in use did not fit the standards agreed for length or for height. They were either Sea-Land's 35-foot containers or, in the Pacific, Matson's 24-foot containers, the product of a rival and parallel project done by a firm shipping pineapples from Hawaii to San Francisco. Eventually, though, the industry settled mainly on 20-foot and 40-foot standard lengths.

McLean's next breakthrough came with the Vietnam War. The United States built up and supplied its troops in Vietnam with constant difficulty, because of the shallow water and inadequacy of the port facilities in Saigon and Da Nang. The military tried again and again to ease the congestion, delay and confusion without much success. It kept getting worse. McLean saw his opportunity and badgered the Pentagon to be allowed to build a container

port at Cam Ranh Bay. He encountered predictable resistance, but his persistence finally paid off in 1967. Sea-Land
constructed a port, at its own risk, and started shipping 600
containers every two weeks. Suddenly, the supply problems of the military were over. Even refrigerated containers
with ice cream aboard joined the rush. Sea-Land garnered
huge sums from the contracts. The restless McLean then
spotted an opportunity to send the empty containers back
via Japan, where they picked up export goods, thus helping
to build the Asian export boom that was to transform the
economies of Japan, Taiwan, Korea, China and eventually
Vietnam. The military soon gave McLean more contracts
for supplying troops in Europe as well, which helped change
the attitude of container-sceptic ports in Europe.

McLean sold Sea-Land to R. J. Reynolds in 1970 and
soon left the company. He tried various other enterprises,
including pig farming and resorts, before buying back
into the shipping industry with the purchase of United
States Lines in 1977. The capacity of container shipping
was growing at 20 per cent a year, and ships were getting
bigger and bigger. Per ton of cargo, a bigger ship cost less to
build, required a smaller crew and consumed less fuel than
a smaller one. The only limit was getting through the locks
of the Panama Canal.

The average speed of container ships dropped during
the 1970s, as fuel costs rose following the oil crises of 1973
and 1978. McLean saw an opportunity to build fourteen
large but slow 'Econships' in South Korea, designed to go
continuously round the world in an easterly direction, thus
avoiding the problem of returning empty. It was a neat idea,
but it did not work. Oil prices fell and round-the-world
timetables proved unreliable. In 1986 McLean Industries

filed for bankruptcy with $1.2bn of debt, the largest bank-
ruptcy in American history at the time. The great risk taker
had taken one risk too many. He was shattered by the
experience and shunned the limelight for a while. He died
in 2001, at the age of eighty-seven, and on the morning of
his funeral container ships all around the world sounded
their whistles at the same time.

The vast container trade across the oceans that today is
vital to the world economy is his legacy. Today, some ships
carry more than 20,000 20-foot containers each; they can
be unloaded and reloaded in just three days. McLean is the
father of modern trade, but invented nothing very novel, let
alone high tech. If he had not made this revolution, somebody else would probably have done it. But he did it.

Was wheeled baggage late?

Having lugged heavy bags through train stations and airports in my youth, I regard the wheeled suitcase as one
of the pinnacles of civilization. But for something so low
tech, it turned up surprisingly late, after the first human
beings had landed on the moon. What was to stop somebody inventing the wheeled suitcase in the 1960s? Why was
it so late in arriving? The wheeled bag seems to be a good
example of tardy innovation that should have happened
sooner. Or does it?

One day in 1970, Bernard Sadow, a senior executive
of a baggage-making company in Massachusetts, went
on holiday in Aruba with his family. On the way back, at
American customs, he joined a queue, periodically picking
up two heavy bags as he shuffled forward. Just then an
airport worker strode past with a piece of heavy machinery

on a wheeled trolley. 'You know, that's what we need for luggage,' Sadow said to his wife. He went home, took four castors off a wardrobe and screwed them on to a suitcase. He then attached a leash to the case and dragged it effortlessly around the house. He applied for a patent on rolling luggage, which was granted in 1972. In the application he wrote: 'The luggage actually glides. Further, substantially any person, regardless of size, strength or age, can easily pull the luggage along without effort or strain.'

But when Sadow took his crude prototype to retailers, one by one they turned him down. The objections were many and varied. Why add the weight of wheels to a suitcase when you could put it on a baggage trolley or hand it to a porter? Why add to the cost? For several years he got nowhere, till eventually Macy's, the department store, commissioned a line of 'bags that glide' from Sadow and the world gradually followed suit.

A glance through the history of patents reveals that Mr Sadow was not the first to try. Arthur Browning had filed a patent on wheeled baggage the year before Sadow, in 1969. Grace and Malcolm McIntyre had tried in 1949. Clarence Norlin had patented suitcases with retractable wheels, the better to fit into spaces, in 1947. Barnett Book filed a patent for a wheeled suitcase in 1945. And Saviour Mastrontonio had patented a 'Luggage Carrier' that could be used to roll 'a bag, satchel, suitcase or the like' in 1925. In the accompanying illustration, a carpet bag rolls in front of a glamorous lady in a stripy dress, propelled by a stiff, long handle.

Clearly, the problem was not a lack of inspiration. Instead, what seems to have stopped wheeled suitcases from catching on was mainly the architecture of stations and airports. Porters were numerous and willing, especially for executives. Platforms and concourses were short and close to drop-off points where cars could drive right up. Staircases abounded. Airports were small. More men than women travelled, and they worried about not seeming strong enough to lift bags. Wheels were heavy, easily broken and apparently with a mind of their own. The reluctant suitcase manufacturers may have been slow to catch on, but they were not all wrong. The rapid expansion of air travel in the 1970s and the increasing distance that passengers had to walk created a tipping point when wheeled suitcases came into their own.

A decade later, Sadow's design was displaced by a superior innovation – the Rollaboard. This was the brainchild of Robert Plath, a pilot with Northwest Airlines. In 1987 he went into his workshop at home and fixed two wheels to one side of the short end of a rectangular suitcase, rather than four to the base of the long side, as Sadow had done. A case could now be tilted and dragged semi-upright with the help of a telescopic handle. Plath sold a few to fellow pilots, but ordinary passengers began to notice them and ask about how to get them, so Plath left the airline and set up Travelpro, which rapidly became a successful business. Four-wheeled versions followed, as well as new aluminium and plastic lightweight versions, and wheels that can roll in any direction so you could push as well as pull. Innovation continues to transform the experience of travel.

The lesson of wheeled baggage is that you often cannot innovate before the world is ready. And that when the world is ready, the idea will be already out there, waiting to be employed: in America, at least. Nothing like this happened in Communist Russia or Mao's China.

Novelty at the table

The restaurant industry is addicted to innovation. It experiences rapid turnover as once-fashionable eating spots give way to new ones, with zero protection from government for those who prefer to resist innovation, zero subsidy for those who wish to innovate and zero overall strategy from experts. It is as close as you can get to a permissionless innovation system. Restaurants must adapt or die. Some can last for decades and become global brands, though even these are constantly having to adjust to shifts in taste. Others are flashes in the pan whose formula catches on briefly, if at all.

Over the past half-century or so, much of the innovation in food has come from importing foreign cooking styles. In 1950, somebody eating out in London would be familiar with French cuisine, but probably not Italian, let alone Indian, Arabic, Japanese, Mexican or Chinese. Today, all these versions of food were on sale in the street market where I bought my (samosa) lunch today, and Korean, Ethiopian, Vietnamese and other styles are not far away. But there is a limit to how many foreign cultures can be found, and this method of innovation will eventually run dry; the restaurant sector has had to get creative in its search for further novelty.

There are occasional new ingredients, though not often: the kiwi fruit and the Chilean sea bass (formerly known as the Patagonian tooth fish) are two examples of foods not eaten before recent decades, but mostly we still eat things like chickens and potatoes in ever more varieties of ways. There are new ways of preparing food, with fancy names like 'foam' or 'jus', and other affectations. There are fusions of styles, with mixed Asian cuisines leading the way. There is the rise of vegan food, with ingenious ways of re-creating the experience of eating red meat (beetroot is key) or fish and chips (banana blossom has surprisingly similar consistency to cod).

In some cases the search for novelty takes on an almost desperate flavour, and goes back to older ingredients or styles. Thus the Danish *haute cuisine* of the chef René Redzepi, whose Copenhagen restaurant Noma won the San Pellegrino award for the most innovative restaurant in the world three years running, starting in 2010, relies at least partly on the retro-novel idea of combining animals with the plants that grow where they live: such as pork neck with bulrushes, violets and malt. Paradoxically, to re-create the extreme localism of ancient hunter-gatherers therefore becomes an innovation.

A study of Noma by two (hungry) professors of innovation stresses that the main method of innovation here is not *de novo* invention, but recombination – bringing old things together in new combinations – and that this is a general feature of innovation elsewhere in the economy: 'Innovation is a process of search and recombination of existing components', a point also made by Joseph Schumpeter in the 1930s: 'Innovation combines components in a new way'.

Can this recombination continue indefinitely? Let's assume there are ten different kinds of meat, ten different kinds of vegetable, ten different kinds of spice or herb, and ten different ways of preparing each. This greatly simplifies the actual situation, but still it results in 10,000 possible different dishes. With a more realistic set of numbers, the number of ways of recombining the ingredients becomes astronomical. So there is not much danger that food will eventually become monotonous and stop changing.

There are even laboratories working on recipes. El Bulli, a restaurant in Spain, was the first to win both a Michelin star and a Pellegrino award. Its owners, Ferran Adrian and Juli Soler, achieved this by investing in their own research and development facility, in which chefs and food scientists develop new recipes during the winter, when the restaurant is closed, for the next year. The Fat Duck, an expensive restaurant in Britain, even developed a seafood dish called Sound of the Sea, with the sound of waves coming from an iPod Nano hidden inside a seashell, after collaborating with psychologists at Oxford University. Those who have studied how chefs innovate report that they follow a process of feed-forward trial and modification, experimenting with variations on a central idea till they hit on a dish that they think will win the approval of customers. It is not very different from the way Thomas Edison improved the light bulb.

But food innovation is not just about ingredients and recipes. It is also about the method of eating. Ray Kroc's realization that simple meals could be prepared to a standard form that could be eaten without plates or forks, and the formula rolled out across the world – McDonald's – is a reminder that it is not the invention but the commercialization that makes the difference. Kroc was a travelling salesman trying to sell mixers for milkshakes against stiff competition. One of his customers was a small chain of Californian hamburger restaurants, run by Richard and Maurice McDonald, that was unusually clean, well organized and popular. 'In my experience, hamburger joints are nothing but jukeboxes, pay phones, smoking rooms, and guys in leather jackets. I wouldn't take my wife to such a place,' he wrote. The McDonald brothers had developed a sort of assembly-line approach to preparing meals that was fast and reliable as long as the menu was simple. Entering into partnership with the brothers, Kroc expanded McDonald's with a franchise model that emphasized uniformity and affordability, while allowing him to keep a tight control of standards, in sharp contrast to the unreliability of fast food in those days. Soon McDonald's was spawning emulators all across America and the world, and eventually its popularity came to earn the snobbish rage of cultural commentators. There can be no greater accolade.

The rise of the sharing economy

It might seem odd to describe the sharing economy as low tech, given its dependence on the internet. But innovations such as eBay, Uber and Airbnb, none of which were foreseen when the internet was launched, are actually simple and non-technical concepts from an earlier era made possible by the connectivity of the modern world. People with spare time can pick up people who need car rides. People with spare rooms can rent them out to people who need somewhere to stay on holiday. People with expertise can lend it to people who need it. People with things to sell find people looking to buy things. These activities were happening before the internet but are becoming much more lucrative and widespread as the world goes online. Not many people saw this coming, though it should have been obvious.

Joe Gebbia and Brian Chesky founded Airbnb in 2008. It has now surpassed five million properties listed in more than 80,000 towns and cities. The gross revenue to renters probably exceeds $40bn a year. These numbers suggest that this innovation fulfils a need. By unlocking the potential value hidden away in people's homes, it brings welcome revenue to the person renting out the property. By supplying more properties to rent it keeps prices lower than they would otherwise be for the person renting. True, it also brings problems, and not just for hotel chains. Cities such as Amsterdam and Dubrovnik have become home-rental monocultures and deserts for permanent residents.

The sharing economy is a form of more from less, or growth by shrinkage – economic enrichment caused by using resources more frugally. In the case of car sharing, many private vehicles stand idle for 95 per cent of their lives; why not use them a bit more? Other examples of the sharing economy are only just getting started. VIPKid, founded in 2013 by Cindy Mi, links up students in China with English-language teachers in America over the internet. By the end of 2018 it was enabling 61,000 teachers to fill their spare time and 500,000 students to learn English. It is sending about $1bn a year from Chinese people to American people. Hipcamp, founded in 2013 by Alyssa Ravasio, enables people who own land near American national parks to find campers willing to pay to pitch a tent on their land. The sharing economy is the oldest idea in the world: connecting people who have more fish than they need with people who have more fruit than they need.

6

Communication and computing

There's a law about Moore's law. The number of people predicting the death of Moore's law doubles every two years.

Peter Lee of Microsoft Research, 2015

The first death of distance

As the three-masted passenger ship *Sully* pitched and rolled in the Atlantic swell, en route from Le Havre to New York in 1832, one night two of the passengers engaged in a momentous conversation after dinner. One was Charles Thomas Jackson of Boston, a geologist and physician and a bit of a genius, though he spent much of his life – before he went mad – furiously claiming priority for other people's scientific discoveries in medicine, geology and technology. He was about to do so now.

The other man was a famous artist, one Samuel Morse. Aged forty-two, he was well regarded by everybody – he had done many portraits, including several of presidents – except himself, who thought he was out of ideas and past his best. He was still trying to finish his masterpiece, on which he had been working for months, a minutely detailed depiction of the Grand Gallery in the Louvre. But the conversation was not about art. According to Morse's recollection five years later, 'We were conversing on the recent scientific discoveries in electro magnetism and the experiments of Ampere.' One of the other passengers inquired whether an electric current could go far down a long wire without being retarded. Jackson replied instantly that Ben Franklin had shown that a current can go as far down a wire as you want and very fast. In that instant, Morse had an idea: perhaps the arrival of the current at the far end of a long wire could somehow bring a message: 'If the presence of electricity can be made visible in any desired part of the circuit I see no reason why intelligence might not be transmitted instantaneously by electricity.' Morse and Jackson then discussed doing experiments to prove this.

Five years later, Morse wrote to the passengers and captain of the *Sully* to get their recollections of that evening. By then he had indeed invented the telegraph, but he was beset by claims from European rivals to have done so before him; he wanted to establish priority. The captain was most helpful: 'I have a distinct remembrance of your suggesting as a thought newly occurred to you, the possibility of a telegraphic communication being effected by electric wires.' So were two passengers. But not Jackson, who now claimed the insight had been his alone: 'I do claim to be the principal in the whole invention made on board the *Sully*. It arose wholly from my materials & was put together at your request by me.' This drove Morse into a fury and eventually to law.

Samuel Morse did more to shrink the world than anybody before or after him. Thanks to his innovations, messages that once took months could now take seconds to reach their destination. Unlike Jackson, Morse did a series of experiments to try to turn the original idea into a device. A suggestion of using relays from Leonard Gale of New York University proved critical, and by 1838 Morse was able to send the message 'A patient waiter is no loser' over a two-mile wire, using a code. In a typical example of simultaneous discovery, he was narrowly beaten to the same goal by two British inventors, Charles Wheatstone and William Cooke, but Morse's version, using a single wire, was better. Moreover, Morse went on to invent a binary digital alphabet to use on the telegraph – Morse code. Like so many inventors he then spent years defending his priority, fighting no fewer than fifteen court actions over his patents: 'I have been so constantly under the necessity of watching the movements of the most unprincipled set of pirates I have ever known, that all my time has been occupied in defense, in putting evidence into something like legal shape that I am the inventor of the Electro-Magnetic Telegraph!' he cried in 1848. He achieved final vindication in the Supreme Court only in 1854.

Morse's real achievement, like that of most innovators, was to battle his way through political and practical obstacles. As his biographer Kenneth Silverman put it:

Morse's claims for himself as an innovator rest most convincingly on the part of his work he valued least, his dogged entrepreneurship. With stubborn longing, he brought his invention into the marketplace despite congressional indifference, frustrating delays, mechanical failures, family troubles, bickering partners, attacks by the press, protracted lawsuits, periods of depression.

In 1843 Congress, after a long siege, appropriated a sum for Morse to install the first telegraph wire, from Washington to Baltimore. The equipment for insulating and entrenching the wire alongside the railway proved hopeless, while his partners proved corrupt and untrustworthy. The next year he changed tack and started suspending the wires from poles, with more success. In May he was able to use the half-completed wire to get news of Henry Clay's nomination for president by the Whig Party convention in Baltimore more than an hour before the train brought confirmation. On 24 May 1844, with the line complete, he transmitted a message all the way from Baltimore to the Supreme Court building in Washington, a quotation from the book of Numbers suggested by Annie Ellsworth, the daughter of a friend: 'what hath God wrought?'

The implications of the telegraph's annihilation of distance were instantly understood in a vast country like America. As an official report put it a few years later:

Doubt has been entertained by many patriotic minds how far the rapid, full, and thorough intercommunication of thought and intelligence, so necessary to a people living under a common representative republic, could be expected to take place throughout such immense bounds. That doubt can no longer exist. It has been resolved and put an end to forever by the triumphant success of the electro-magnetic telegraph of Professor Morse.

Telegraph wires soon criss-crossed the continents, with 42,000 miles laid by 1855 in America alone. In 1850 the first underwater cable was dropped across the English Channel, wrapped in 'gutta percha', an insulator derived from the rubber tree. A transatlantic cable followed in 1866 and a submarine cable from Britain to India in 1870, reaching Australia in 1872. Because of its overseas empire, Britain dominated the marine cable-laying industry and London lay at the hub of a web of submarine cables. Submarine cable capacity increased tenfold in the thirty years from 1870.

There was widespread utopian hope about the telegraph's impact on society, as there would be 150 years later for the internet. The wires would make war less likely, keep families in touch, transform the practice of finance and deter crime, commentators speculated. One newspaper, the *Utica Gazette*, waxed lyrical: 'fly, you tyrants, assassins and thieves, you haters of light, law, and liberty, for the telegraph is at your heels.'

Once the telegraph was in use, the telephone was bound to follow at some point. In 1876, in what is often cited as a spectacular case of simultaneous invention, Alexander Graham Bell arrived at the patent office to file a patent on the invention of the telephone, and just two hours later Elisha Gray arrived at the same patent office with an application for the very same thing. In fact the two had been rivals in the race to develop a telephone (or harmonic telegraph, as they called it) for several years, and there was ample evidence they were snooping on each other's work and each other's negotiations with the patent office. So this is one of those cases where the coincidence is not uncanny, just competitive.

In fact, we now know that both Bell and Gray were beaten to the telephone by Antonio Meucci, an Italian who emigrated to Cuba, then New York. He was experimenting with 'a vibrating diaphragm and an electrified magnet', the key ingredients of the telephone receiver, back in 1857 and filed a patent caveat in 1871. He built lots of devices and even used them to communicate between floors in his house in Staten Island. The reason history forgot Meucci is because, unlike the determined Bell, he raised no money to develop the idea or defend his patents, and his candle factory went broke, leaving him in poverty and bankruptcy. He was an inventor, but not an innovator.

The miracle of wireless

Guglielmo Marconi is unusual among innovators in several respects. First, he was upper class, using his butler as a research assistant in his laboratory in a family villa. Second, he was good at both the technical invention and the commercial production of his new idea, becoming a leading businessman. And third, he really did get some of his ideas from science, from experiments by Heinrich Hertz, whereas most inventors before that date were engineers or technologists but not scientists. But in one respect Marconi was entirely typical: he did a huge amount of trial and error.

Marconi was born in an apartment in a palace in Bologna and raised at first in a hilltop villa outside the city. He was the son of a wealthy Italian businessman and an Irish mother from the Jameson whiskey-distilling family. His family moved to Bedford in England for four years, then to Florence and then Livorno, where the young Marconi was privately tutored in science. His cousin Daisy

Prescott remembered that he was always inventing things as a boy and was obsessed with electricity, both parents encouraging his hobby.

In 1888 Heinrich Hertz published the results of ingenious experiments demonstrating the existence of electromagnetic waves, propagating at the speed of light, as predicted by the physicist James Clerk Maxwell. 'We just have these mysterious electromagnetic waves that we cannot see with the naked eye. But they are there,' he wrote. But as for applications: 'Nothing, I guess.'

Marconi read about this and began to think there might be applications, in wireless telegraphy, to signal Morse messages without cables. There were already several ideas about how to do this over very short distances, using electrical induction in the ground, water or air, but none had proved practical. There were also claims to have broadcast signals before Marconi, without a full understanding of how, most prominently by an American dentist named Mahlon Loomis, who in 1872 patented the 'aerial telegraph', using a kite to produce 'a disturbance in the electrical equilibrium of the atmosphere'. He even got Congress to vote a large sum for its development but it went nowhere.

Quite when and how Marconi did his first experiment is uncertain, because his own later accounts kept changing as he reinvented bits of his biography. But there is little doubt that by the end of 1895, at the Villa Griffone, he had sent a signal of three taps across the hillside to a receiver, where his assistant fired a gun to acknowledge receipt. Marconi, just twenty-two years old, promptly moved to London to apply for a British patent on his invention, which he was convinced would make him a fortune. In London he was assisted by his cousin Mary Coleridge, great-niece of Samuel Taylor Coleridge, the author of *The Ancient Mariner*, and herself a notable writer. Mary introduced Marconi to her close friend Henry Newbolt, then a prominent lawyer and later to become a pillar of the literary and political establishment as an author of patriotic poems. Newbolt immediately realized the promise of the invention and that the contract Marconi was being offered by an interested firm was much to his disadvantage. He advised him to seek an expert patent lawyer and secured, through his own social connections, an introduction to Alan Campbell Swinton, later president of the Wireless Society, who in his turn introduced Marconi to William Preece of the Post Office, which was trying to develop communications between lightships. Of course, it helped that Marconi was respectable, with well-connected family in London, but these people did not have to help him. They did so because they saw a fruitful possibility and wished it well. Like the telegraph's pioneers half a century before, and the internet's pioneers a century later, Marconi believed that freeing up global communication could only enhance peace and harmony among peoples. This utopianism was catching. The physicist Sir William Crookes had also foreseen the use of Hertzian waves to transmit information – it fitted with his belief in psychic forces – and he once wrote of using them in 'improving harvests, killing parasites, purifying sewage, eliminating diseases and controlling weather'.

Had Marconi not lived, radio would still have come to life in the 1890s. Others such as Jagadish Chandra Bose in India, Oliver Lodge in Britain and Alexander Popov in Russia were doing and publishing experiments that used electromagnetic waves to create action at a distance, though not always for communication. Some, such as

Édouard Branly in France and Augusto Righi in Bologna, were inventing better devices for transmitting and receiving such waves. And then there was Nikola Tesla, the restless genius and inventor of the electric motor, the alternating current and lots of ideas relating to radio. Marconi was just another experimenter, though a very good one, but thanks to Newbolt he was quick to patent what he had found as broadly as possible, thus demonstrating that it is the system of intellectual property that contributes to the singling out of individual inventors, as much as the other way around.

Marconi also knew how to take the devices and ideas of others and put them together in simple and practical form. As his biographer Marcus Raboy puts it: 'working by trial and error over several months in 1895, Marconi perfected the coherer, invented a stable tapper, increased the efficiency of the induction coil, connected a Morse inker and telegraphic relay to the transmitter and receiver, and controlled the resulting electrical sparks.'

He was also more commercially minded than most of his rivals. In 1897 he transmitted signals across nine miles of water in the Bristol Channel and established stations on the Isle of Wight and in Bournemouth to continue developing and demonstrating the technology. By 1899 he had transmitted a message across the English Channel and by 1902 across the Atlantic from Cape Breton in Canada to Poldhu in Cornwall (his claim to have heard a transatlantic transmission in 1901 with weaker receivers was probably true, because it might have bounced off the ionosphere, then unknown, but was widely disbelieved at the time). Within a few years he was embroiled in exhausting legal battles, especially with the American inventors Reginald Fessenden and Lee de Forest. History records that all of them made key improvements in radio, crucial to turning it into a voice system, rather than a Morse system, and that the expensive argument in court was a waste of time.

Marconi was slow to see the role that broadcasting would play in the story of radio, thinking of it more as a communication medium. But by the 1920s the possibilities of broadcasting were undeniable. 'For the first time in the history of the world man is now able to appeal by means of direct speech to a million of his followers, and there is nothing to prevent an appeal being made to fifty millions of men and women at the same time,' wrote Marconi, perhaps beginning to see that his invention had a dark side too. On 12 February 1931, at Marconi's side, the Pope launched Vatican radio in a blaze of global publicity. At a reception afterwards, the Pope thanked both Marconi and God for putting 'such a miraculous instrument as wireless at the service of humanity'.

Others of a less benign intent took notice of the Vatican example. 'It would not have been possible for us to take power or to use it in the ways we have without the radio,' noted Josef Goebbels in August 1933. A detailed analysis by a group of economists in 2013 shows that in the elections of September 1930 the Nazi vote share rose less in areas where radios were more numerous, because broadcasts generally had a mild anti-Nazi slant. Heavy pro-Nazi propaganda began on the radio immediately after Adolf Hitler became chancellor in January 1933, and only five weeks later, in the last proper elections, radio's impact was reversed: the Nazi vote share increased more in places where more people had access to radios. (A similar pattern was observed in the Rwanda genocide of 1993: the more people in an area who had access to the 'hate radio station' RTLM, the greater the violence against Tutsis.)

The Nazis used radio massively to influence Austrians and Sudeten Germans as well as domestically. They developed a cheap radio receiver, the *Volksempfänger*, or people's radio, costing 76 Reichsmarks, specially to ensure that they could reach more people. 'All Germany hears the Führer with the People's Radio,' boasted a poster advertising it in 1936. Oswald Mosley tried, via his wife, to get Hitler's support to broadcast to Britain from Germany. Even in the democracies, where Father Charles Coughlin was using radio to foment anger against bankers and Jews among his 30 million listeners, while Franklin Roosevelt was using it to sell his policies, the impact of radio on the polarization of society was huge – shades of what has happened more recently with social media. 'Have I done the world a good, or have I added a menace?' Marconi asked, in 1934. Five years earlier Mussolini had made Marconi a marquis.

For reasons that are not entirely clear, network television had the opposite effect of radio, bringing people back towards a social consensus, sometimes stiflingly so, rather than polarizing them. If there was a moment that encapsulates this shift, it was in April 1954, when the American people got their first glimpse of Senator Joe McCarthy via television. They did not like what they saw and McCarthy's bubble burst immediately. 'The American people have had a look at you for six weeks. You are not fooling anyone,' said Senator Stuart Symington shortly afterwards. It was this centripetal effect that has gone into reverse with the arrival of social media, I think, a polarizing force like early radio.

Who invented the computer?

If the origin of the steam engine is lost in the fog of the early 1700s, when obscure and impoverished men worked without much reward and nobody chronicled their adventures, then how much easier it will be to decide who invented the computer, an innovation of the mid-twentieth century with all the leading players amply supplied with opportunities to record their work for posterity and everybody aware they were making history. Yet we have no such luck. The origination of the computer is as mysterious and confusing as that of far more ancient and uncertain innovations. There is nobody who deserves the accolade of the inventor of the computer. There is instead a regiment of people who made crucial contributions to a process that was so incremental and gradual, cross-fertilized and networked, that there is no moment or place where it can be argued that the computer came into existence, any more than there is a moment when a child becomes an adult.

The computer, as we know it, has four indispensable ingredients to distinguish it from a mere calculator. It must be digital (in particular binary), electronic, programmable and general purpose – that is, capable of carrying out any logical task, at least in principle. In addition, it must actually work. After an exhaustive survey of many claims, the historian Walter Isaacson concludes that the first machine to meet all these criteria is the ENIAC, the Electronic Numerical Integrator and Computer, which began operating towards the end of 1945 at the University of Pennsylvania. Weighing 30 tons, and the size of a small house, containing over 17,000 vacuum tubes, the ENIAC worked successfully for many years and was the design copied by most computers immediately thereafter. The ENIAC was the brainchild of three people, a cerebral physicist named John Mauchly, a perfectionist engineer named Presper Eckert and an efficient soldier named Herman Goldstine.

But to pull this machine out and imply that its construction marks a sudden break with the past when the world had no computers would be a big mistake. For a start, the ENIAC was not binary, but decimal. And Mauchly ended up losing a lengthy and bitter legal dispute trying to defend his patent on the ENIAC's design. The judge ruled that he had stolen a lot of the key ideas from an obscure experimental machine built in Iowa after a blinding insight in a roadside bar in 1937 by a talented engineer named John Vincent Atanasoff. Yet Atanasoff's machine was small, was not fully electronic, never worked and was not programmable or general purpose, so the outcome of the lawsuit makes little sense – except to lawyers. True, Mauchly got some good ideas from his visit to Iowa to see Atanasoff, but then that's the way innovation works.

A better candidate to challenge the ENIAC's claim might be Colossus, the computer built at Bletchley Park in Britain to crack German codes. Colossus preceded the ENIAC by almost two years, the first version being finished in December 1943 and the second, larger version going operational in June 1944: within a few weeks it had decoded some of Hitler's orders in the battle for Normandy. Colossus was fully electronic, digital (and binary, unlike the ENIAC) and programmable. But it was designed as a single- not a general-purpose machine. Besides, even in the 1970s its story was still shrouded in secrecy, so its influence on later machines was smaller. There again, even if you give Colossus its due, to whom should we give the credit for designing it? The construction was led largely by an engineer named Tommy Flowers, a pioneer of using vacuum tubes in complex telephone circuits, and his boss was the mathematician Max Newman, but they consulted Alan Turing, the tortured code-breaking genius of Bletchley, who had already built 200 electromechanical devices called Bombes in Hut 8. After the war ended, Frederic Williams's 'Manchester Baby' computer started working at Manchester University in June 1948, influenced by Tommy Flowers and Alan Turing. It qualifies as the world's first stored-program, electronic computer – the first von Neumann architecture – further complicating the picture. And its offspring, the Manchester Mark 1 was developed into the first commercially available computer, the Ferranti Mark 1.

But mention of Turing reminds us that the idea of a general-purpose computer should perhaps be the thing we celebrate rather than an actual machine. Turing's remarkable mathematical paper 'On Computable Numbers', published in 1937, was the first logical demonstration that a universal computer, capable of doing any logical task, could exist. Today we call such things 'Turing machines'. At Princeton, in 1937, Turing actually built a machine that used electrical relay switches to turn letters into binary numbers for encoding. Perhaps that deserves to be designated the eureka moment, even though it was neither completed, nor a computer.

Yet Turing's ideas were ethereal and mathematical. More practical was the precocious Master's thesis of Claude Shannon, an MIT student who worked at Bell Labs, also in the summer of 1937. Shannon pointed out that Boolean algebra, developed nearly a century before by the mathematician George Boole, could be instantiated in electrical circuits. The word 'and' could be two switches in sequence, 'or' could be two switches in parallel, and so on. 'It is possible to perform complex mathematical operations by means of relay circuits,' he concluded. Shannon's paper was later dubbed the 'Magna Carta of the Information Age' by *Scientific American*.

And when discussing the theory behind computers, one cannot leave out Johnny von Neumann, the hyper-intelligent and gregarious Hungarian, whose name is forever attached to the architecture of the modern computer, and who was Turing's mentor at Princeton. In June 1945 von Neumann wrote the most influential guide to the structure of computers, called rather obscurely 'First Draft of a Report on the EDVAC', in which he set out for the first time the notion that an all-purpose computer should have programs stored in its memory alongside the data. As influential documents go, this one was pivotal, though it was unfinished and had been written by hand mostly on trains. (The Electronic Discrete Variable Automatic Computer, or EDVAC, was the successor to the ENIAC, completed in 1949.)

Yes, but hang on, where did von Neumann get his ideas from in the 'First Draft'? Largely from studying the Mark 1 computer at Harvard, which was built by a team led by a professor-turned-naval-officer by the name of Howard Aiken. The Mark 1 was not electronic, so cannot claim the prize itself, but it was programmable, more so than the ENIAC. It preceded the ENIAC by two years, and was programmed with punched tape, a crucial innovation in itself. Herman Goldstine bumped into von Neumann on a railway station platform at Aberdeen, Maryland, in August 1944 and told him about the ENIAC. Von Neumann arranged to come and see it and quickly realized that he was looking at something that was much quicker than the Mark 1 at doing a calculation but was much slower and more cumbersome to re-program. Hence his suggestion that the EDVAC be designed to hold programs stored inside it alongside data. Thus von Neumann's unique privilege in being able to travel freely between the teams (and his high security clearance) made him a vital cross-pollinator of ideas.

However, IBM then disputed Aiken's claim to have designed the Mark 1, arguing that its engineers had responded to Aiken's commission by developing a series of small but crucial inventions to improve and refine the Mark 1, and which Aiken had no hand in. This is a reminder that IBM was already not just in existence but dominating a massive industry producing calculating machines for human 'computers'. IBM had been formed in 1924 from the merger of various other companies, one of which had been founded to help tabulate the 1890 US census. One of the tributary streams of the computer therefore comes from this industry, a point often forgotten by those who like to see innovation starting with professors, rather than business people.

Moreover, von Neumann's 'First Draft' paper heavily drew upon, perhaps even plagiarized, the thinking and writing of Aiken's deputy, the formidably talented Grace Hopper. Given that Hopper deserves much credit for the idea of program subroutines, as well as the compiler, she is arguably the mother of the software industry, surely just as important an innovation as the hardware of computers. Later, she invented natural-language programming, another seminal breakthrough. So perhaps the more important origin of the computer lies in this software story rather than the hardware one. And yet Hopper must share a chunk of the credit with the programmers of the ENIAC, also women, who pioneered the writing of programs, because the ENIAC had been originally expected to be used to make firing tables for the trajectories of artillery shells in different atmospheric conditions, a task no longer needed so urgently after 1945. One of those ENIAC pioneers, Jean Jennings, astutely observed that they only got their chance because the men in charge thought the reconfiguration of a computer was a menial task: 'If the ENIAC's administrators had known how crucial programming would be to the functioning of the electronic computer, and how complex it would prove to be, they might have been more hesitant to give such an important role to women.'

Ah, but in rescuing Hopper and Jennings from a man-dominated hardware story, it is necessary to go further back and recognize their precursors. Almost exactly the same relationship between a male hardware pioneer and a female software pioneer, echoing the Aiken–Hopper relationship, had cropped up a century before, in the 1840s. Far ahead of his time, an inventor by the name of Charles Babbage had started to build two mechanical calculators, the first of which, the Difference Engine, designed to solve differential equations, was supported by the British government to the tune of £17,000, an enormous sum. The second, the Analytical Engine, was in essence going to be a general-purpose computer, but Babbage never finished it. The concept was enough, though, to inspire the spectacular mind of Ada Byron, Countess of Lovelace, to write a series of notes in which she prefigured many of the concepts of modern computers, including software and subroutines. She realized that computers could handle any subjects, not just numbers, she saw that data could be represented in digital form, and she published what was in effect the first computer program. If anybody was a genius far ahead of her time in this story it was probably her.

Yet Babbage and Lovelace, too, must be placed in context. They knew that the Jacquard loom, already in use in the textile industry, was a sort of program: a set of cards that automatically lifted the weave in the right sequence to produce a particular pattern in a cloth. Just because this was the province of journeymen industrialists rather than gentleman-philosophers, it must not be omitted from the story. Note that Ada Lovelace, who gave due credit to the Jacquard loom, and celebrated it, found herself on the opposite side from her father of a now-familiar debate, between pro-technology and anti-technology arguments. He, the poet Lord Byron, made a passionate speech in the House of Lords defending the Luddites who had been smashing such looms on the grounds that automation destroyed jobs. His daughter was all for innovation.

In summary, the ENIAC was not so much invented as evolved through the combination and adaptation of precursor ideas and machines. And it was only a stage in the gradual evolution of the computer. If there was an annus mirabilis of the computer, when these cross-fertilizations of ideas and devices happened most fruitfully, it was, Walter Isaacson thinks, 1937. It was in that year that Turing published 'On Computable Numbers', that Claude Shannon explained how circuits of switches could embody Boolean algebra, that George Stibitz at Bell Labs proposed an electrical calculator, that Howard Aiken commissioned the Mark 1, and that John Vincent Atanasoff conceived key features of an electronic computer. Also in 1937, Konrad Zuse in Berlin built a prototype of a calculator that could

read a program from a punched tape. His Z3 machine, finished in May 1941 in Berlin, can lay claim to having been
an all-purpose, programmable digital computer – as early
as any other.

By then of course his country was at war. The development of the computer is always supposed to have been
accelerated by wartime funding, but the counterfactual of
what would have happened if war had not broken out (in
1939 for Britain and Germany, in 1941 for America), is hard
to discern. By 1945, without war, there would undoubtedly
have been devices that were electronic, digital, programmable and general purpose. Indeed, without the need for
secrecy, they might have evolved faster, as separate teams
shared ideas faster and used their devices for other purposes than calculating the trajectories of artillery shells or
decoding the secret messages of enemies. Had Zuse, Turing,
von Neumann, Mauchly, Hopper and Aiken all met at a
conference in peacetime, who knows what would have
happened and how fast?

The ever-shrinking transistor

Innovators are often unreasonable people: restless, quar-
relsome, unsatisfied and ambitious. Often, they are
immigrants, especially on the West Coast of America. Not
always, though. Sometimes they can be quiet, unassuming,
modest and sensible stay-at-home types. The person whose
career and insights best capture the extraordinary evolution
of the computer between 1950 and 2000 was one such.
Gordon Moore was at the centre of the industry throughout this period and he understood and explained better
than most that it was an evolution, not a revolution. Apart

from graduate school at Caltech and a couple of unhappy years out East, he barely left the Bay Area, let alone California. Unusually for a Californian, he was a native, who grew up in the small town of Pescadero on the Pacific coast just over the hills from what is now called Silicon Valley, going to San Jose State College for undergraduate studies. There he met and married a fellow student, Betty Whitaker.

As a child, Moore had been taciturn to the point that his teachers worried about it. Throughout his life he left it to partners like his colleague Andy Grove, or his wife, Betty, to fight his battles for him. 'He was either constitutionally unable or simply unwilling to do what a manager has to do,' said Grove, a man toughened by surviving both Nazi and Communist regimes in his native Hungary. Moore's chief recreation was fishing, a pastime that requires patience above all else. And unlike some entrepreneurs he was – and is, now in his nineties – just plain nice, according to almost everybody who knows him. His self-effacing nature somehow captures the point that innovation in computers was and is not really a story of heroic inventors making sudden breakthroughs, but an incremental, inexorable inevitable progression driven by the needs of what Kevin Kelly calls 'the technium' itself. More so than flamboyant figures like Steve Jobs, who managed to make a personality cult in a revolution that was not really about personalities.

In 1965 Moore was asked by an industry magazine called *Electronics* to write an article about the future. He was then at Fairchild Semiconductor, having been one of the 'Traitorous Eight' who defected from the firm run by the dictatorial and irascible William Shockley to set up their own company six years before, where they had invented the integrated circuit of miniature transistors printed on a silicon chip. Moore and Robert Noyce would defect again to set up Intel in 1968. In the 1965 article Moore predicted that miniaturization of electronics would continue and that it would one day deliver 'such wonders as home computers . . . automatic controls for automobiles, and personal portable communications equipment'. But that prescient remark is not why the article deserves a special place in history. It was this paragraph that gave Gordon Moore, like Boyle and Hooke and Ohm, his own scientific law:

> The complexity for minimum component costs has increased at a rate of roughly a factor of two per year. Certainly over the short term this rate can be expected to continue, if not to increase. Over the longer term, the rate of increase is a bit more uncertain, although there is no reason to believe it will not remain nearly constant for at least 10 years.

Moore was effectively forecasting the steady but rapid progress of miniaturization and cost reduction, doubling every year, through a virtuous circle in which cheaper circuits led to new uses, which would lead to more investment, which would lead to cheaper microchips for the same output of power. The unique feature of this technology is that a smaller transistor not only uses less power and generates less heat, but can be switched on and off faster, so it works better and is more reliable. The faster and cheaper chips got, the more uses they found. Moore's colleague Robert Noyce deliberately underpriced microchips, so that more people would use them in more applications, growing the market.

By 1975 the number of components on a chip had passed 65,000, just as Moore had forecast, and it kept on growing as the size of each transistor shrank and shrank, though in that year Moore revised his estimate of the rate of change to doubling the number of transistors on a chip every two years. By then Moore was chief executive of Intel and was presiding over its explosive growth and the transition to making microprocessors, rather than memory chips: essentially programmable computers on single silicon chips. Calculations by Moore's friend and champion, Carver Mead, showed that there was a long way to go before miniaturization hit a limit.

Moore's Law kept on going not just for ten years but for about fifty years, to everybody's surprise. Yet it probably has now at last run out of steam. The atomic limit is in sight. Transistors have shrunk to less than 100 atoms across, and there are billions on each chip. Since there are now trillions of chips in existence, that means there are billions of trillions of transistors on Planet Earth. They are probably now within an order of magnitude of equalling the number of grains of sand on the planet. Most sand grains, like most microchips, are made largely of silicon, albeit in oxidized form. But whereas sand grains have random – and therefore probable – structures, silicon chips have highly non-random, and therefore improbable, structures.

Looking back over the half-century since Moore first spotted his Law, what is remarkable is how steady the progression was. There was no acceleration, there were no dips and pauses, no echoes of what was happening in the rest of the world, no leaps as a result of breakthrough inventions. Wars and recessions, booms and discoveries, seemed to have no impact on Moore's Law. Also, as Ray Kurzweil was to point out later, Moore's Law in silicon turned out to be a progression, not a leap, from the vacuum tubes and mechanical relays of previous years: the number of switches delivered for a given cost in a computer trundled upwards, showing no sign of sudden breakthrough when the transistor was invented, or the integrated circuit. Most surprising of all, discovering Moore's Law had no effect on Moore's Law. Knowing that the cost of a given amount of processing power would halve in two years ought surely to have been valuable information, allowing an enterprising innovator to jump ahead and achieve that goal now. Yet it never happened. Why not? Mainly because it took each incremental stage to work out how to get to the next stage.

This was encapsulated in Intel's famous 'tick-tock' corporate strategy: tick was the release of a new chip every other year, tock was the fine-tuning of the design in the intervening years, preparatory to the next launch. But there was also a degree of self-fulfilling prophecy about Moore's Law. It became a prescription for, not a description of, what was happening in the industry. Gordon Moore, speaking in 1976, put it this way:

> This is the heart of the cost reduction machine that the semiconductor industry has developed. We put a product of given complexity into production; we work on refining the process, eliminating the defects. We gradually move the yield to higher and higher levels. Then we design a still more complex product utilizing all of the improvements, and put that into production. The complexity of our product grows exponentially with time.

Silicon chips alone could not bring a computer revolution. For that, there needed to be new computer designs, new software and new uses. Throughout the 1960s and 1970s, as Moore foresaw, there was a symbiotic relationship between hardware and software, as there had been between cars and oil. Each industry fed the other with innovative demand and innovative supply. Yet even as the technology went global, more and more the digital industry became concentrated in Silicon Valley, a name coined in 1971, for reasons of historical accident: Stanford University's aggressive pursuit of defence research dollars led it to spawn a lot of electronics startups, and those startups gave birth to others, which spawned still others. Yet the role of academia in this story was surprisingly small. Though it educated many of the pioneers of the digital explosion in physics or electrical engineering, and though of course there was basic physics underlying many of the technologies, neither hardware nor software followed a simple route from pure science to applied.

Companies as well as people were drawn to the west side of the San Francisco Bay to seize opportunities, catch talent and eavesdrop on the industry leaders. As the biologist and former vice-chancellor of Buckingham University, Terence Kealey, has argued, innovation can be like a club: you pay your dues and get access to its facilities. The corporate culture that developed in the Bay Area was egalitarian and open: in most firms, starting with Intel, executives had no reserved parking spaces, large offices or hierarchical ranks, and they encouraged the free exchange of ideas sometimes to the point of chaos. Intellectual property hardly mattered in the digital industry: there was not usually time to get or defend a patent before the next advance overtook it. Competition was ruthless and incessant, but so were collaboration and cross-pollination.

The innovations came rolling off the silicon, digital production line: the microprocessor in 1971, the first video games in 1972, the TCP/IP protocols that made the internet possible in 1973, the Xerox Parc Alto computer with its graphical user interface in 1974, Steve Jobs's and Steve Wozniak's Apple 1 in 1975, the Cray 1 supercomputer in 1976, the Atari video game console in 1977, the laser disc in 1978, the 'worm', ancestor of the first computer viruses, in 1979, the Sinclair ZX80 hobbyist computer in 1980, the IBM PC in 1981, Lotus 123 software in 1982, the CD-ROM in 1983, the word 'cyberspace' in 1984, Stewart Brand's Whole Earth 'Lectronic Link (Well) in 1985, the Connexion machine in 1986, the GSM standard for mobile phones in 1987, Stephen Wolfram's Mathematica language in 1988, Nintendo's Game Boy and Toshiba's Dynabook in 1989, the World Wide Web in 1990, Linus Torvald's Linux in 1991, the film *Terminator 2* in 1992, Intel's Pentium processor in 1993, the zip disc in 1994, Windows 95 in 1995, the Palm Pilot in 1996, the defeat of the world chess champion, Garry Kasparov, by IBM's Deep Blue in 1997, Apple's colourful iMac in 1998, Nvidia's consumer graphics processing unit, the GEForce 256, in 1999, the Sims in 2000. And on and on and on.

It became routine and unexceptional to expect radical innovations every few months, an unprecedented state of affairs in the history of humanity. Almost anybody could be an innovator, because thanks to the inexorable logic unleashed and identified by Gordon Moore and his friends, the new was almost always automatically cheaper and faster than the old. So invention meant innovation too.

Not that every idea worked. There were plenty of dead ends along the way. Interactive television. Fifth-generation computing. Parallel processing. Virtual reality. Artificial intelligence. At various times each of these phrases was popular with governments and in the media, and each attracted vast sums of money, but proved premature or exaggerated. The technology and culture of computing were advancing by trial and error on a massive and widespread scale, in hardware, software and consumer products. Looking back, history endows the triers who made the fewest errors with the soubriquet of genius, but for the most part they were lucky to have tried the right thing at the right time. Gates, Jobs, Brin, Page, Bezos, Zuckerberg were all products of the technium's advance, as much as they were causes. In this most egalitarian of industries, with its invention of the sharing economy, a surprising number of billionaires emerged.

Again and again people were caught out by the speed of the fall in cost of computing and communicating, leaving future commentators with a rich seam of embarrassing quotations to mine. Often it was those closest to the industry about to be disrupted who least saw it coming. Thomas Watson, the head of IBM, said in 1943 that 'there is a world market for maybe five computers.' Tunis Craven, commissioner of the Federal Communications Commission, said in 1961: 'there is practically no chance communications space satellites will be used to provide better telephone, telegraph, television or radio service inside the United States.' Marty Cooper, who has as good a claim as anybody to have invented the mobile phone, or cell phone, said, while director of research at Motorola in 1981: 'Cellular phones will absolutely not replace local wire systems. Even if you project it beyond our lifetimes, it won't be cheap enough.' Tim Harford points out that in the futuristic film *Blade Runner*, made in 1982, robots are so life-like that a policeman falls in love with one, but to ask her out, he calls her from a payphone, not a mobile.

The surprise of search engines and social media

I use search engines every day. I can no longer imagine life without them. How on earth did we manage to track down the information we needed? I use them to seek out news, facts, people, products, entertainment, train times, weather, ideas and practical advice. They have changed the world as surely as steam engines did. In instances where they are not available, like finding a real book on a real shelf in my house, I find myself yearning for them. They may not be the most sophisticated or difficult of software tools, but they are certainly the most lucrative. Search is probably worth nearly a trillion dollars a year and has eaten the revenue of much of the media, as well as enabled the growth of online retail. Search engines, I venture to suggest, are a big part of what the internet delivers to people in real life – that and social media.

I use social media every day too, to keep in touch with friends, family and what people are saying about the news and each other. Hardly an unmixed blessing, but it is hard to remember life without it. How on earth did we manage to meet up, to stay in touch or to know what was going on? In the second decade of the twenty-first century social media exploded into the biggest and second most lucrative use of the internet and is changing the course of politics and society.

Yet here is a paradox. There is an inevitability about both search engines and social media. If Larry Page had never met Sergei Brin, if Mark Zuckerberg had not got into Harvard, then we would still have search engines and social media. Both already existed when they started Google and Facebook. Yet before search engines or social media existed, I don't think anybody forecast that they would exist, let alone grow so vast, certainly not in any detail. Something can be inevitable in retrospect, and entirely mysterious in prospect. This asymmetry of innovation is surprising.

The developments of the search engine and social media follow the usual path of innovation: incremental, gradual, serendipitous and inexorable; few eureka moments or sudden breakthroughs. You can choose to go right back to the posse of MIT defence-contracting academics, such as Vannevar Bush and J. C. R. Licklider, in the post-war period, writing about the coming networks of computers and hinting at the idea of new forms of indexing and net-working. Here is Bush in 1945: 'The summation of human experience is being expanded at a prodigious rate, and the means we use for threading through the consequent maze to the momentarily important item is the same as was used in the days of square-rigged ships.' And here is Licklider in his influential essay, written in 1964, on 'Libraries of the Future', imagining a future in which, over the weekend, a computer replies to a detailed question: 'Over the week-end it retrieved over 10,000 documents, scanned them all for sections rich in relevant material, analyzed all the rich sections into statements in a high-order predicate calculus, and entered the statements into the data base of the question-answering subsystem.' But frankly such prehistory tells you only how little they foresaw instant search of millions of sources. A series of developments in the field of computer software made the internet possible, which made the search engine inevitable: time sharing, packet switching, the World Wide Web and more. Then in 1990 the very first recognizable search engine appeared, though inevitably there are rivals to the title.

Its name was Archie, and it was the brainchild of Alan Emtage, a student at McGill University in Montreal and two of his colleagues. This was before the World Wide Web was in public use and Archie used the FTP protocol. By 1993 Archie was commercialized and growing fast. Its speed was variable: 'While it responds in seconds on a Saturday night, it can take five minutes to several hours to answer simple queries during a weekday afternoon.' Emtage never patented it and never made a cent.

By 1994 Webcrawler and Lycos were setting the pace with their new text-crawling bots, gathering links and key words to index and dump in databases. These were soon followed by Altavista, Excite and Yahoo!. Search engines were entering their promiscuous phase, with many different options for users. Yet still nobody saw what was coming. Those closest to the front still expected people to wander into the internet and stumble across things, rather than arrive with specific goals in mind. 'The shift from exploration and discovery to the intent-based search of today was inconceivable,' said Srinija Srinivasan, Yahoo!'s first editor-in-chief.

Then Larry met Sergey. Taking part in an orientation programme before joining graduate school at Stanford, a university addicted by then to spinning out tech companies, Larry Page found himself guided by a young student named Sergey Brin. 'We both found each other obnoxious,' said Brin later. Both were second-generation academics in technology. Page's parents were academic computer scientists in

Michigan; Brin's were a mathematician and an engineer in Moscow, then Maryland. Both young men had been steeped in computer talk, and hobbyist computers, since childhood.

Page began to study the links between web pages, with a view to ranking them by popularity, and had the idea, reportedly after waking from a dream in the night, of cataloguing every link on the exponentially expanding web. He created a web crawler to go from link to link, and soon had a database that ate up half of Stanford's internet bandwidth. But the purpose was annotating the web, not searching it. 'Amazingly, I had no thought of building a search engine. The idea wasn't even on the radar,' Page said. That asymmetry again.

By now Brin had brought his mathematical expertise and his effervescent personality to Page's project, named BackRub, then Page Rank, and finally Google, a misspelled word for a big number that worked well as a verb. When they began to use it for search, they realized they had a much more intelligent engine than anything on the market, because it ranked sites that the world thought were important enough to link to higher than those that happened to contain key words. Page discovered that three of the four biggest search engines could not even find themselves online. As Walter Isaacson has argued:

> Their approach was in fact a melding of machine and human intelligence. Their algorithm relied on the billions of human judgments made by people when they created links from their own websites. It was an automated way to tap into the wisdom of humans – in other words, a higher form of human–computer symbiosis.

Bit by bit, they tweaked the programmes till they got better results. Both Page and Brin wanted to start a proper business, not just invent something that others would profit from, but Stanford insisted they publish, so in 1998 they produced their now famous paper 'The Anatomy of a Large-Scale Hypertextual Web Search Engine', which began: 'In this paper, we present Google . . .' With eager backing from venture capitalists they set up in a garage and began to build a business. Only later were they persuaded by the venture capitalist Andy Bechtolsheim to make advertising the central generator of revenue.

As with search engines, so social media took the world by surprise. I recall reviewing two books in the 1990s that forecast gloomily that the internet was going to make people antisocial. We were going to retreat into our bedrooms and play games, starting a spiral of social degradation of apocalyptic proportions. In fact, within a decade, the internet was being used for rampant social engagement on a massive scale. Today, teachers and parents worry about the incessant online social distraction that keeps children from studying, not to mention the risk of cyber-bullying and peer pressure.

Facebook launched in February 2004 as a Harvard University networking site. Mark Zuckerberg had been employed by two fellow students, Cameron and Tyler Winklevoss, the previous November, to program a social-networking site called Harvard Connection, but had then developed his own version instead, called 'the facebook', got the financial backing of Eduardo Saverin and later Sean Parker and Peter Thiel, and took the idea commercial. The Winklevosses had a point when they sued him, but in the Wild West of digital innovation it was first past the post.

Social media took the world by surprise in another way too. Far from ushering in an era of utopian democratic enlightenment in which the world is flat, everybody is sharing and we all see each other's point of view, it plunged us into a maze of echo chambers and filter bubbles in which we spend our time confirming our biases and railing against the opinions of others. It polarized, enraged, depressed, addicted and soured us.

Aza Raskin, who was one of the inventors of the 'infinite scroll', by which we can just keep rolling through our social media feeds for ever, now regrets that he did this. He says it was one of the first features of technology designed 'not to help you but to keep you'. He now works to try to redirect the tech industry towards more beneficial and less addictive results. There seems little doubt that any information technology, when young, can have strong and unhelpful effects, but that it usually gets tamed. This was true of printing, cheap newspapers and radio.

Eli Pariser, in a 2011 book called *The Filter Bubble*, locates two key moments in time when this echo effect took hold. One was on 4 December 2009, when Google announced that it was personalizing its search results, based on signals from users' habits and preferences. Different people would (and do) get different results when searching for the same term. Pariser cites the case of two friends, both left-leaning East Coast women, who searched for the term BP at the height of the news about an oil spill in the Gulf of Mexico. One got environmental news, the other investment advice.

The second event was four months later, when Facebook launched Facebook Everywhere, which allowed users to 'like' anything they found on the web, so that everything could be personalized: news, advertising, information, whatever. The personalization revolution was key to the rise of Amazon, too. From the start, when it was just an online bookstore, Amazon used a new technique called collaborative filtering to customize its search results, albeit clumsily at first.

The harvesting of personal data and preferences to allow personalization still seemed innocent back then, and Barack Obama was praised for his use of targeting with social media in the 2012 election, but in the years since the mood has changed. There is little doubt that filter bubbles and cable television are responsible for political polarization all around the world, with left-leaning people moving left and right-leaning people moving right – and with sinister, state-backed forces in Russia and elsewhere encouraging the trend. In a recent study, a group of social scientists paid a large sample of Democrats and Republicans who visit Twitter at least three times a week to spend a month following a bot that passed on messages from the opposite political ideology. They found that Republicans became even more conservative after following a left-leaning twitter bot, Democrats slightly more liberal after following a conservative Twitter bot.

As Pariser predicted: 'Left to their own devices, personalization filters serve up a kind of invisible autopropaganda, indoctrinating us with our own ideas, amplifying our desire for things that are familiar and leaving us oblivious to the dangers lurking in the dark territory of the unknown.' Innovation often takes the world in surprising directions.

We have been here before. The invention of printing caused political and social upheaval in Western societies that polarized society and killed a lot of people, mainly in wars fought about whether the body of Christ was literally or figuratively present at the Eucharist and whether the Pope was infallible. It also ushered in an enlightenment of knowledge and reason unprecedented in scope and depth. The combination of printing press, paper and movable type, brought together by Johann Gutenberg around 1450, was an information innovation that caused huge social change, little of it predicted and not all of it good. As Steven Johnson has observed, Gutenberg's press was 'a classic combinatorial innovation, more bricolage than breakthrough', each of its elements having been already invented by others, including those who operated wine presses. But even if you call Gutenberg the inventor, Martin Luther was the true innovator, transforming the use of printing from an obscure business confined mainly to the ecclesiastical elite to a mass-market operation aimed at ordinary people. He produced short, readable pamphlets in German rather than Latin. By 1519 he had published forty-five works in almost 300 editions and was Europe's most published author. Like Jeff Bezos at Amazon or Mark Zuckerberg at Facebook, he had realized the potential of a new technology on a huge scale.

Machines that learn

Today, artificial intelligence is the trendiest frontier of the information world. It is also one of the oldest ideas in computing and with a history of chronic and repeated failure to deliver. At Dartmouth College in 1956 – that is, more than sixty years ago – John McCarthy and Marvin Minsky organized a conference on artificial intelligence and launched the phrase on what would prove to be a gullible world. McCarthy thought that 'if a carefully selected group of scientists work on it together for a summer', they might make significant progress towards a thinking computer, while a breakthrough that would match or replicate human intelligence inside a computer was about twenty years away. It did not happen, funders lost patience and research in the field of clever computers went into an 'AI winter'. Something similar happened again in the 1980s. As Walter Isaacson drily remarks: 'decade after decade, new waves of experts have claimed that artificial intelligence was on the visible horizon, perhaps only twenty years away. Yet it has remained a mirage, always about twenty years away.'

The problem is partly that when computers learn clever tricks, we tend immediately to reclassify the task as non-intelligent, realizing that it is achieved without under-standing. The anticipation of your desires that smartphones offer in everyday life is artificially intelligent, but we don't think of it like that, because we know it's just an unthink-ing algorithm. When the IBM computer Deep Blue narrowly defeated Gary Kasparov in a chess match in 1997, in a mile-stone of computer cleverness, the achievement was dismissed as a triumph for brute force. Deep Blue evaluated 330 mil-lion positions a second, but did it think, imagine or feel?

Twenty years later, in 2016, a London startup firm called Deep Mind stunned the world when its program AlphaGo defeated the world champion Lee Sedol at the game of Go, in a tournament televised all across Asia. The episode marked a turning point in the artificial-intelligence story, with a new wave of excitement, not least in China. Will this turn into another 'AI winter' or is this time different?

Go is too complex a game for the brute-force tech-niques that helped Deep Blue, and the crucial ingredient of

AlphaGo is its ability to learn. It was not taught the rules of Go, but intuited them from examples of games using neural networks (the latest version of the program does not consult human games at all). Thus the human beings who programmed AlphaGo have no idea why it chose the moves it did. Move 37 in Game 2 was described by an expert as 'creative' and 'unique', because it broke all the normal rules and seemed to be stupid. Lee Sedol took an unusually long time to respond, and although his counter was also unusually clever, he eventually lost the game following a series of equally brilliant moves by AlphaGo at moves 151, 157 and 159.

Thus the focus of artificial intelligence has shifted from the 'expert system' approach in which clever people try to impart their knowledge to computers, to a learning approach in which programs find ways to solve problems themselves. This was made possible by three features of the modern computing world: new software, new hardware and new data. The new software is the brainchild at least partly of Geoffrey Hinton, a Toronto-based scientist of British extraction. Hinton's family tree is studded with famous mathematicians, entomologists and economists, and he himself trained as a psychologist before developing in the early 1990s the notion of 'back propagation' in neural networks. This is essentially a feedback method that enables such networks to make internal representations of the world through 'unsupervised learning'. Such programs were very limited in their capacities until the deluge of data increased exponentially in the last decade but are now remarkably good at inducing generalizations and insights from deep draughts of data, without being instructed in how to do so in any detail. Thus, for example, by absorbing lots of examples of prostate tumour scans, a computer can now learn how to identify and delineate a tumour for targeted radiology, a task that takes a long time to do by hand when done by a highly paid radiologist.

However, new hardware was also crucial to this change, and it came from a surprising source: the computer games industry. The central feature of a computer is the CPU, or central processing unit. This includes one or a few 'cores', which do the calculations, and lots of cache memory. For most tasks this is fine, but the games industry found that in creating realistic, apparently three-dimensional images, it needed a different type of chip: one with hundreds of cores that can handle hundreds of software threads at a time. This 'graphics processing unit', or GPU, does not replace the CPU but it does augment it, and has proved invaluable for making deep learning, through back propagation, possible. Nvidia, a manufacturer of graphics cards, coined the term GPU when it launched the GeForce 256 graphics card in 1999, still aimed wholly at gaming. The company had been founded in 1993 by Jensen Huang, a Taiwanese who immigrated to Oregon as a child, and two colleagues. They were not the inventors of parallel chips, but the improvers. It was not until 2007 that the first general-purpose GPU hit the market. By 2018 Nvidia was unveiling robots that could learn to do tasks simply by observing human beings.

So the AI breakthroughs of recent years are the product of new tools as much as new data and new ideas. There will be significant teething problems before machine learning can be trusted by ordinary people in everyday life. A team of scientists at the University of Washington in Seattle trained a neural network to distinguish images of husky dogs from images of wolves, just by giving it twenty example photographs. However, they had deliberately selected only pictures of wolves against snowy backgrounds; and only dogs against grassy backgrounds. Sure enough, it turned out that the algorithm was paying much more attention to the background than the animal. When people were asked if they trusted this neural network to make good decisions, they were much less likely to say yes after this fact had been explained to them. Thus, the explainability – the opportunity to interrogate an algorithm as to its reasoning – will be a key ingredient of making artificial intelligence trustworthy. In another case, Amazon found that a neural network intended to help with hiring had begun to discriminate against women. Yet human brains are also black boxes whose reasoning is sometimes obscure, so we may be holding machines to a higher standard than people.

For the moment, the safest bet is that artificial intelligence will augment rather than replace people, as automation has done for centuries. Even in the case of chess playing, the most successful teams these days are 'centaurs', that is to say combinations of algorithms and people. The same will undoubtedly be true of driving. I already rely on my car to warn me when a car is passing me in an outside lane, or when a car is approaching as I back out of a parking space. Many more such 'intelligent' tricks will be at my disposal in the future, but the day when I settle into the car, tell it where to take me and go to sleep at the wheel is – in my opinion – a fair way off.

7

Prehistoric innovation

There is no great invention, from fire to flying,
that has not been hailed as an insult to some god.
J. B. S. HALDANE

The first farmers

Before the last two centuries, innovation was rare. A person could live his or her whole life without once experiencing a new technology: carts, ploughs, axes, candles, creeds and corn looked the same when you died as when you were born. Innovation happened but sporadically and slowly. Travel still further back and the rate of change slows down even more. As the dial on your time machine reaches 10,000 years before today, you would alight in a world that was changing so slowly as to be imperceptible not just in one lifetime but in ten. And yet you would be landing in the middle of one of the most momentous innovations of all: the adoption of farming.

Farming changed the human being from a sparse population of predators and gatherers into a landscape-altering, high-density, ecosystem changer. Valleys like the Nile, the Indus, the Euphrates, the Ganges and the Yangtze became largely man-made ecosystems, in which specialized grasses put human beings to work tending them and planting them, while the steppes and hills of Asia became dominated by cattle and sheep and horses that employed people to protect and tend them. Nomads settled; population density leapt upwards, to be checked only by outbreaks of new diseases or famines. Soon strange new cultural innovations like kings, gods and wars began to dominate events. Farming was an innovation as vast in its implications as the steam engine or the computer. Like the Industrial Revolution, the agricultural one was all about energy: producing more of it, in more concentrated form, and directing it towards the reversal of entropy through the creation of more human bodies at the expense of other species. Any book about how innovation works must tackle this ancient innovation.

Alas I cannot resort to biography to tell the tale, though it is a fair bet that the farming 'revolution' had its prehistoric Norman Borlaugs. But in other ways the invention of agriculture will throw up some familiar patterns. First, there is the phenomenon of simultaneous invention. Just as the light bulb appeared independently in the 1870s in many different parts of the world around the same time, so agriculture did the same. Admittedly the 'same time' might in this case span a millennium or two, but the point is that compared with the half a million years or more that our species had been hunting and gathering, a few thousand years is but an eye blink. In that time people took up farming in at least six different places wholly independent of each other: in the Near East, China, Africa, South America, North and Central America, and New Guinea. There is no evidence that any of these people got the idea of farming from each other and the particulars of crop and cultivation were different in each case. The wheat farmers of Mesopotamia did not influence the millet farmers of China, let alone the potato farmers of the Andes or the yam farmers of New Guinea.

This coincidence implies either that human brains had evolved in parallel towards the capability to have the idea of farming, which seems unlikely, or that there was something new about the conditions of the time that made farming likely. There was indeed something special: the climate. Prior to 12,000 years ago the world was in a deep ice age. This meant that it was much cooler, with huge ice sheets covering much of Europe and North America, as well as mountainous regions further south. But it also meant that the world was much drier, because colder oceans evaporated less moisture so rainfall was sparser and lighter.

Africa was plagued by prolonged droughts in which desert conditions lasted for decades; Lake Victoria dried out altogether 16,000 years ago and the Kalahari desert was larger and drier. The Amazon rain forest had shrunk to a quilt of isolated patches of forest interspersed with grasslands. Huge dust clouds blew about the world, staining the ice sheets of Antarctica. With so much moisture locked up in ice caps, sea level was hundreds of feet lower than today. With the seas so cold and well stratified, carbon dioxide had dissolved in the water so that there were just 190 parts per million left in the atmosphere at the last glacial maximum, or under 0.02 per cent. This made it very difficult for plants to grow fast, if at all, especially in arid areas, because plants lose moisture when they open their pores to absorb carbon dioxide. Experiments show that at 190 ppm plants like wheat and rice would yield only about one-third as much grain as today even when well supplied with water and nutrients.

The correlation between dust in the ice core records from Antarctica and very low carbon dioxide levels is strong, meaning that plants had retreated from many mountainous and arid areas, leaving dusty, unstable soil. Rainfall was rare. Around 20,000 years ago, to judge by the Antarctic ice cores, the dust storms must have been truly terrible, darkening the skies almost everywhere in the world for weeks at a time. At that time, around a hundred times as much dust was deposited in Antarctica as in the interglacial warm period 10,000 years later. This was not a good time for a large-brain, small-gut, energy-intensive ape to try to live off a diet of plants on any continent. Better to let the sparse herds of specialized grazing animals – horses, bison, antelopes and deer – gather what calories they could into concentrated lumps of meat, and eat them. In some places there might be tubers to dig up – a human speciality – or nuts to harvest, but domesticating these was not going to be easy for another reason: the climate was extremely volatile.

None of this was known until recent years, when good ice cores, especially from Antarctica and Greenland, became available. Various records suggest that the temperature was far more volatile at the last glacial maximum than today, both in the polar regions and in the tropics. Global temperatures were four times as variable, from one decade to the next, as today. Mediterranean pollen records, for example, show wild fluctuations during the ice age, compared with more recent millennia. This would have made farming impossible. Droughts or long cold spells would have forced farmers to migrate, leaving any shrivelled crops behind. The balance of incentives favoured nomadic hunter-gathering.

In 2001 two pioneers in the study of cultural evolution, Pete Richerson and Rob Boyd, published a seminal paper that argued for the first time that agriculture was 'impossible during the Pleistocene [ice age] but mandatory during the Holocene [current interglacial]'. Almost as soon as the climate changed to warmer, wetter and more stable conditions, with higher carbon dioxide levels, people began shifting to more plant-intensive diets and to making ecosystems more intensively productive of human food. 'Almost all trajectories of subsistence intensification in the Holocene are progressive,' they wrote, 'and eventually agriculture became the dominant strategy in all but the marginal environments.' In that sense farming was compulsory, unavoidable, which is why it happened in so many different places.

In the archaeological record, farming can look sudden, when a settlement is replaced by another with farmed grains, but a closer look at archaeological sites preserved in the Sea of Galilee reveals a much more gradual pattern, in which hunter-gatherers fed on fish and gazelles for thousands of years while very slowly increasing their seasonal reliance on grass seeds cut from the surrounding land in autumn. To start with the intervention would have looked like gardening. People must have sometimes saved seed and scattered it on wet ground in spring to encourage more growth, while chasing away birds, weeds and grazing animals to protect it. Perhaps they did this on silty islands in rivers, which were fertile but seedless. The person who picked up some particularly heavy seeds from a fortuitous hybrid of emmer and einkorn, two early ancestors of wheat, might not have done so on purpose. The resulting hybrid – bread wheat – is a hexaploid genetic monster with heavy seeds incapable of dispersing and surviving without human intervention. Bit by bit, the seeds would have responded through natural selection: heavier, free-threshing seeds, more easily harvested, would have come to be more densely represented in the seed kept to be sown. A virtuous circle would have come into being. In a sense the plant took the initiative.

There is another similarity with later bursts of innovation: it happened at a time of plenty and in a place of plenty. Just as innovation flourishes in wealthy, growing and well-connected places at a time of peace and relative prosperity – California today, Newcastle in Stephenson's day, Renaissance Italy in Fibonacci's day – so farming began in the warm, well-watered valleys of the Euphrates, the Yangtze and the Mississippi, or in the rich and sun-baked soils of New Guinea and the Andes. The shift to farming was not a sign of desperation any more than the invention of the computer was. True, a life of farming proved often to be one of drudgery and malnutrition for the poorest, but this was because the poorest were not dead: in hunter-gathering societies those at the margins of society, or unfit because of injury or disease, simply died. Farming kept people alive long enough to raise offspring even if they were poor. There are parallels here too with modern innovation. Computers enable people to hold down good jobs who would have struggled in the heavy industries of the Victorian era.

By taking up farming, human beings not only changed the genes of wheat plants and cows. They changed their own genes too. A further innovation very clearly shows how genes and culture co-evolve. This was dairy farming, first invented around 8,000 years ago. People had domesticated cattle by now and had begun to milk them. They encountered a problem: although cow's milk was excellent food for infants, adult human beings were, like all adult mammals, unable to digest lactose, the chief sugar in milk. The lactase gene was programmed to switch off at weaning, when no longer needed. Milk was still a deliciously nutritious drink for adults, full of proteins and fats, but the lactose could not be broken down, so people would have found raw milk an uncomfortable, flatulence-inducing food – as many people descended from non-dairy cultures do today. Better to make cheese with it, where the lactose was digested first by bacteria.

But one day a mutant person was born whose lactase gene failed to switch off at weaning. He or she enjoyed much more reward from drinking milk and grew strong and healthy, bearing more children than other people. His or her genes came to dominate the population. These 'lactose-tolerant' mutations came to prominence in several different parts of the world, in Eurasia and Africa, always in close coincidence with the invention of dairy farming. But it was clearly the dairy farming that led to the selection of the genetic change, not vice versa. The genetic innovation in people was an inevitable consequence of the cultural innovation.

The invention of the dog

Long before farming was invented, human beings made a crucial innovation that transformed their fortunes: the dog. It was the first animal to be domesticated and become an ecological companion of people all over the world, hunting alongside them to their mutual benefit, before later being selected for a huge variety of specialized roles. Who made this innovation, how and where? The domestication of the dog happened in Eurasia. We know this because dogs are most closely related to Eurasian wolves and were domesticated before people moved into the Americas. Dogs made it to Australia with one wave of people, probably not the first wave, and re-wilded themselves as dingoes.

The date of dog domestication has recently been pinned down a little more closely, thanks to genetics, and it is more ancient than anybody expected. Krishna Veeramah, a geneticist at Stony Brook University in New York, analysed DNA from three ancient remains of dogs, dating from 4,700–7,000 years ago, and compared the results with DNA sequences from 5,649 modern dogs and wolves. His team's conclusion, published in 2017, is that dogs diverged from wolves around 40,000 years ago, and subsequently split into two (eastern and western) branches of the dog family tree around 20,000 years ago: Chinese village dogs are genetically distinct from European breeds after this date. This suggests that the domestication happened only once, and that it was between 20,000 and 40,000 years ago. Also that it might have happened in western Europe, or south-east Asia, or somewhere in between.

The DNA from a wolf that died 35,000 years ago in northern Siberia had already hinted that by then wolves were separate from dogs. Thus well before the last glacial maximum, but during a much colder period than today, people living on the Eurasian mainland somehow made friends with wild wolves and turned them into useful tools.

Or was it the other way around? Human beings are just as useful to dogs as vice versa, I sometimes reflect as my dog snoozes while I write books to be able to afford to buy it food and a bed to lie on. It's fairly likely that the domestication began with wolves tentatively hanging around human camps to try to scavenge leftover carcasses. The bolder ones risked being speared, but got more food; gradually boldness in the presence of people became commoner in one group of wolves till people saw the advantage of having semi-tame wolves hanging around, perhaps because they provided an early-warning system of an attack or perhaps because they tracked down wounded prey animals.

A fascinating long experiment conducted in Siberia since the 1960s shows how this would work and reveals something surprising about the evolution of tameness in both dogs and people. The experiment concerned foxes, but its point is more general. In 1937 Nikolai Belyaev, a well-known geneticist, was arrested and executed without trial for showing an unhealthy interest in Western genetic science. His brother Dmitri was just twenty at the time, but he went on to become a geneticist, too, though he was very careful to pay lip service to the prevailing environmentalist dogma of Stalinism. He went to work in a laboratory that studied fur-bearing animals, and in 1958 he moved to Novosibirsk to join the Institute of Cytology and Genetics of the Siberian Department of the Soviet Academy of Sciences. Here he decided to study silver foxes.

The silver fox is a subspecies of the red fox, originally from Canada, that was being farmed in Siberia for its fur. But farmed is perhaps the wrong word, because these were wild animals kept in cages. They showed little sign of adapting to human captivity and becoming tame. According to the Lysenko dogma then promoted by Nikita Khrushchev, captivity should itself cause tameness to appear, but it clearly was not happening. Belyaev decided to try selective breeding instead. He did this by a very simple trick. He bred from the least frightened foxes in every generation:

the ones that snarled least when their cage was approached. Then he did the same for fox pups, selecting the friendliest, least timid and least aggressive pups. Out of a thousand pups bred every year, 200 were chosen as parents for the next generation. This went on for half a century.

Almost immediately, the researchers noticed a difference. In the fourth generation some of the pups approached people spontaneously, wagging their tails – something wild foxes do not do. Within a few more generations, Belyaev had enthusiastically tame foxes that rushed over to lick their human friends. What was most surprising, though, was that the foxes had also changed in appearance. They had curly tails, floppy ears, slightly more feminine heads and white patches on their foreheads – as are often found in domesticated cattle, horses and other pets. They also had larger litters and began breeding at younger ages and out of season. Belyaev repeated the experiment with mink and rats, with similar results.

It turned out that in selecting for docility, Belyaev had also selected for genetic mutations that came with other traits: a domestication syndrome. In particular, he had unwittingly promoted a delay in the migration of the animal's 'neural crest' cells during development. These cells disperse throughout the embryo and give rise to certain tissues within organs such as the skin and the brain. Most of the cells that produce black pigment derive from the neural crest, and it is the paucity of such cells in the head that give domesticated animals their white blaze on the face. In Belyaev's foxes, the melanoblasts that make fur dark are delayed in their migration and the result is white patches in the fur. Delayed neural crest cells are responsible too for floppy ears and smaller jaws.

Richard Wrangham, an anthropologist at Harvard University, hypothesizes that neural crest cells are also crucial to those parts of the brain that regulate stress, fear and aggression. The effect is to make an individual animal less likely to indulge in reactive aggression. And this, Wrangham points out, is a peculiar feature of human beings as well as their pets. Unlike, say chimpanzees, we can shuffle on to a crowded bus without killing each other, something chimps would find impossible. We are just as good if not better at planned aggression, but not reactive aggression. The same is true of dogs. A wolf, or a chimp, is a dangerous pet because though it may be friendly for years it can suddenly react with lethal violence if touched in the wrong way: Wrangham recounts the experience of somebody who tried to pat a captive wolf as you would a dog and nearly lost an arm.

Human beings are rarely like this. From birth we are amazingly tolerant of other people. It looks like we too are a domesticated species, selected by a bunch of Dr Belyaevs – each other – to be less reactively aggressive to strangers, the better to survive in urban, agricultural or dense hunter-gatherer settlements. At some point in human prehistory we must have weeded out people who had fast-migrating neural crest cells and hair-trigger reactions. Whether we did this by executing them, generation after generation, or by ostracizing them, or by sending them into battle, or some combination of the three, we continued doing so into more recent history, and the penal system does so even today.

We gave ourselves a big dose of the domestication syndrome, compared with our ape-man ancestors: more feminine features, smaller jaws, resulting in more crowded teeth, less in the way of sex differences, more in the way of continuous sexual activity and even sometimes a blaze of white hair on the front of the head. Smaller brains too: ancient skeletons show that human brains have shrunk by some 20 per cent in the last 20,000 years, a fact that has often puzzled biologists. Brains shrink during domestication in others species, too, including dogs. Wrangham writes: 'The differences between modern humans and our earlier ancestors have a clear pattern. They look like the differences between a dog and a wolf.' There is even now evidence of which genes were changed to achieve this result. For instance, the BRAF gene shows strong recent evolutionary selection in cats, horses and people and this relates to neural crest cell migration.

Perhaps it is stretching it to call domestication genetics an innovation, though the dog itself was certainly a great invention. But is it really so different from the Industrial Revolution, which was not very deliberate and whose impact was barely noted at the time either? Innovation is a lot less directed and planned, even today, than we tend to think. Most innovation consists of the non-random retention of variations in design.

The (Stone Age) great leap forward

If the invention of agriculture and dogs seems impossibly ancient and slow as innovations go, how much more distant is the invention of sophisticated tools in the later part of the Stone Age, at least 100,000 years ago – known as the human revolution. Yet that too was a burst of innovation and was driven by the same sort of forces as gave us container ships and mobile phones, albeit much more slowly.

Before the human revolution, ape-men did have tools. Two million years ago our hominid ancestors had technology to go with their big brains. They knapped flint rocks to make sharp-edged 'axes', with which to butcher meat or process materials. But for a very long time they did not have innovation, at least not in any sense we would recognize. Artefacts stayed the same for hundreds of thousands of years, and so did the methods for making them. They also looked the same thousands of miles apart in different continents, and perhaps even in different species of ape-men: it's hard to tell *Homo erectus* tools from others. This remains a baffling phenomenon that scientists struggle to explain. Technology existed without even a hint of cultural diversity, let alone a pattern of innovation.

Perhaps a parallel with birds' nests is helpful. These are artefacts, technologies even, built by brain-equipped vertebrates capable of learning. Yet the structure of a bird's nest and the materials it is built from are characteristic of each species and vary little over thousands of miles or from decade to decade. Swallows make mud cups, wrens make moss balls, pigeons make stick platforms. Nest building is an innate instinct, which is why it varies so little. Perhaps toolmaking was an innate instinct in *Homo erectus*.

Meir Finkel and Ran Barkal of Tel Aviv University believe that this conservatism may be confined to stone tools but not to other hominid technologies and habits. The Acheulean hand-axe, in particular, seems to have become fixed in cultural conformity more than other tools. This is the tear-drop-shaped, sharp-edged stone tool with which *Homo erectus* carved up the carcasses of large mammals. It is sometimes found discarded next to the bones of horses or rhinos. They write: 'It is our contention that the handaxe's role in Acheulean adaptation was pivotal and it thus became fixed in human society, probably through the psychological bias towards majority imitation, which subsequently became a social norm or tradition.'

But then gradually innovation began to stir. By 160,000 years ago in Africa new tool kits were beginning to appear. Complicated recipes, like heat-treating stone tools, emerged. By 45,000 years ago in the Middle East an explosion of novel tools was evident, and the acceleration itself accelerated over the coming millennia, resulting in boomerangs and bows and arrows. Just as with the Industrial Revolution, so the human revolution has proved to be a mirage. Europe experienced a surge of new stone technologies at around 45,000 years ago but only – we now know – because it was experiencing 'catch-up growth', like South Korea rapidly industrializing after the Second World War. In the case of Europe, it was catching up with Africa, where new technologies had been emerging much more gradually but much earlier. This was first pointed out by two anthropologists, Sally McBrearty and Alison Brooks, in 2000, in a scathing criticism of the theory that the human revolution in Europe implied a change in the working of the human brain: 'This view of events stems from a profound Eurocentric bias and a failure to appreciate the depth and breadth of the African archaeological record.' Many of the components of the human revolution emerged tens of thousands of years earlier in Africa, including smaller stone tools and blades, bone tools and the movement of tools over long distances, probably through trade.

Yes, but why Africa and why then? Searching for the origin of this slow burst of Stone Age innovation takes us to southern Africa, and one set of caves in particular.

Pinnacle Point is on the south-east coast of South Africa and lies in a place where the worst of the desert conditions did not apply during ice ages: this coast would have remained fairly lush even when the Kalahari desert expanded and grew extremely dry. In those days, with sea level much lower, the caves were higher above the sea, but close enough to the coast to be used as a shelter by early people who left behind the remains of seafood meals and tools. For that reason, the archaeologist Curtis Marean selected the caves for special attention some years ago. He found evidence of human occupation extending back at least 160,000 years, even beyond the last interglacial period before the present one. And he found evidence of complex human behaviour starting many tens of millennia earlier than expected: a varied mix of specialized stone tools, the use of colouring materials, the use of fire to harden tools and so forth: the sort of thing that only turns up elsewhere much later. He also found extensive evidence of people eating seafood.

One example is the use of 'microliths'. These are small flakes of stone chipped off large blocks, then shaped and hardened by fire to make them into lethal tips of project-ile weapons. Marean finds them in cave deposits from 71,000 years ago at Pinnacle Point. He cannot rule out that they were used to make arrows, which would imply the invention of the bow many thousands of years earlier than previously thought, or spear throwers, also thought to be a later invention. 'Early modern humans in South Africa had the cognition to design and transmit at high fidelity these complex recipe technologies,' Marean concludes. Micro-liths enabled people to kill animals at greater distance, with less risk of injury, and also to eliminate enemies before they could get within range of throwing their spears by hand. This might have included the Neanderthals when the technology reached Europe. But why were these southern Africans so innovative?

Here is what Marean thinks was going on at Pinnacle Point. Elsewhere in Africa, food is always and everywhere unpredictable or thinly spread. Trees that produce fruit or nuts, or herds that migrate through after the rains, can provide sudden rich bounty, but it does not last. By contrast, things that stick around, like tubers and small antelope, are sparsely distributed. The life of a hunter-gatherer, therefore, must be mobile, nomadic and lonely: bands would be small and distances between bands large. In such conditions, the collective brain is small – there is not much room for specialization or a division of labour. Hunter-gatherers in such habitats retain very simple tools, cultures and habits.

In just a few places on the African continent, however, resources are rich, predictable and persistent. Some lakes might be like this, if you know how to catch fish, kill crocodiles and hippos or bring down birds. Coasts would be good, too, but not everywhere. Tropical beaches and rocky shores are comparatively unproductive. As is the Mediterranean littoral with its small tides and weak currents. But the coast of south Africa, where cool, nutrient-rich waters bring abundant fish, seals and shellfish, would have provided a buffet of unusually reliable richness. Marean thinks this launched human society into innovation by making it dense, sedentary and territorial. Entering the coastal foraging niche enabled people to settle down in fairly large aggregations to defend their particular section of coast. Living in 'villages' replete with stored food, costly material culture and a concentration of offspring made people targets for rivals to raid, incentivizing the invention of the spear thrower or bow. In ants, the emergence of highly social behaviour with divisions of labour also coincides with the invention of a fixed nest. In effect, to put it ana-chronistically, the first bow maker may have had the time to experiment because his friends were catching enough fish to 'pay' him to do research as part of a 'defence budget'.

Note once again the correlation of innovation with richness. Just as innovation thrives today in the wealthy Silicon Valley, and thrived in the rich Italian city states of the Renaissance, and in the Greek or Chinese city states of ancient times, and just as farming was invented among the plenty of fertile river valleys, so Stone Age innovation began alongside the bounty of seafood.

Mark Thomas and his colleagues at University College London wrote a paper in 2009 arguing that innovation in the Upper Paleolithic is all about demography. Dense populations inevitably spur human technological change, because they create the conditions in which people can specialize. The most striking evidence for this idea comes from Tasmania and concerns, not innovation, but 'disinnov-ation'. The Tasmanian people became isolated 10,000 years ago when rising sea levels at the end of the ice age cut the island off from mainland Australia. They remained effect-ively uncontacted until Western explorers arrived. During these millennia of isolation, the population of the island was small – about 4,000 people – and not only showed little sign of technological innovation, but actually gave up some of the technologies that were there at the start. By the end, Tasmanians had no bone tools, no cold-weather clothing, no hafted tools, no nets, no barbed spears, fishing spears, spear throwers or boomerangs. In a key paper pub-lished in 2004, the anthropologist Joe Henrich explained this by reference to the sudden reduction in the 'effective population size' upon isolation. The Tasmanians went from being a small part of a huge population to the whole of a small population. That meant they could no longer draw upon the ideas and discoveries of many people. Given the need to learn skills, the technology shrank to what could be supported by limited specialization within a small population.

The startling idea here (as I argued in *The Rational Optimist*) is that some time before 150,000 years ago human beings had become reliant on a collective, social brain mediated through specialization and exchange. If you cut people off from exchange, you lower their chances of innovating. This notion gets support from other lines of evidence. Pacific islanders have more complicated fishing technologies if they live on larger islands and – crucially – if those islands have good trading links with other islands. Modern human hunter-gatherers arriving in Europe were able to get objects from a long way away, through trade, in sharp contrast to Neanderthals, who used only local materials and apparently did not have trade among strangers. If they could get objects from a long distance then they could get ideas too. And to this day small, isolated populations show simplified technology and slow rates of innovation. The Andaman Islands are an example among hunter-gatherers, while the North Koreans are an example among industrial people.

More recent history teaches the same lesson. Innovation flourished in cities that traded freely with other cities, in India, China, Phoenicia, Greece, Arabia, Italy, Holland and Britain: places where ideas could meet and mate to produce new ideas. Innovation is a collective phenomenon that happens between, not within, brains. Therein lies a lesson for the modern world.

The feast made possible by fire

Innovations like steam and social media change culture. Fire was an innovation that went one step further and changed human anatomy. Nobody yet knows for sure when fire was invented or where. According to hints in the archaeological evidence it could have been half a million years ago or two million and it could have happened once or many times. But the anatomical evidence is rather stronger: human beings cannot subsist on raw food; their bodies are adapted to cooked food and probably have been so for almost two million years. That implies controlled fire.

Some people try to live off raw food today, and the result is that they always lose weight, and suffer from infertility and chronic energy deficiency, however much they fill their bellies with nuts and fruits. A German study of more than 500 raw-food faddists, who ate most of their food raw, concluded that 'a strict raw food diet cannot guarantee an adequate energy supply.' And this was among people who were eating domesticated and easily digested fruits and vegetables rather than wild food, let alone trekking through forests energetically looking for food, as their chimpanzee equivalents would do while thriving on such diets. Most raw foodists have to include some cooked food in their diets. The human gut is just not adapted to extract enough energy from raw vegetables, raw meat, raw nuts or raw fruit. Which is very odd, when you think about it, because it is not true of any other species, including domesticated ones like dogs.

Every human society that has been contacted cooks food, however simple their ecosystem and their dependence on particular species: from the Inuit to the Sentinelese to the Fuegians. Every hunter-gatherer society revolves around the cooking fire. They might snack on raw food during the day but they return to a camp fire to cook the evening meal. Richard Wrangham recounts the case of Dougal Robertson's family, who survived thirty-one days at sea on a life raft eating their fill of turtles and fish. They lived but lost a lot of weight and fantasized about cooked meals. People are also far more susceptible to stomach bugs from rotten meat, and to bitter and toxic compounds in wild plants, than other apes. We really are adapted to cooking food.

Cooking predigests food. It gelatinizes starch, almost doubling its digestible energy. It denatures proteins, increasing the energy available from eating an egg or a steak by 40 per cent or so. It is like having an external extra stomach. Cooking therefore explains why we have small teeth, small stomachs and a gut that is only a little over half as big as in other apes, relative to our body weight. This small gut costs us less to run – 10 per cent less energy is burned by people just keeping the alimentary canal alive, compared with other apes. So the cooking fire not only provides us with energy, but also saves us energy. As Leslie Aiello argued, this was a crucial step in the expansion of the human brain. In adding to the size of an energy-hungry organ atop the neck, early hominids could not sacrifice the liver or the muscles, but they could and did save on the stomach and gut. Cooking therefore released the possibility of bigger brains.

The shift to a larger brain and a smaller gut seems to have happened a little after two million years ago when *Homo habilis* was replaced by *Homo erectus* in Africa and elsewhere in the world, though these are two perhaps misleadingly precise labels for a gradual and piecemeal change over a long period and with a sparse fossil record. Till recently, the change was explained by a shift to meat eating. But Richard Wrangham in his book *Catching Fire* argues that this cannot make sense, because the human gut is ill equipped for digesting raw meat, compared with say a dog, and heavily dependent on either fat (in cold climates) or carbohydrate (in warm ones) to balance the meat we eat. He therefore argues that it was cooking that explains the change: the emergence of *Homo erectus* saw smaller teeth, a narrower pelvis and a less flared rib cage – all implying a smaller gut. Plus a big increase in brain volume.

Not all are convinced by this idea. In particular, the evidence suggests there was no sudden step change in brain size, only a gradual increase over time. A bit like Moore's Law in the twentieth century, when there were changes in technology, but still a gradual increase in computing power for a given price, so in the hominin fossil record there looks to be a gradual, steady increase in brain size, despite a series of discontinuous species.

How would a *Homo erectus* have invented cooking? Fire was not unknown of course. Indeed, at certain seasons it must have been a common occurrence to see lightning-ignited grass fires. Chimpanzees take this natural phenomenon in their stride. Did *Homo erectus* perhaps get into the habit of hanging around such fires and catching small animals that rushed to escape the flames; or to looking for the charred bodies of creatures that got caught in the fire, having found that they tasted good and made a satisfying meal: lizards, rodents, birds' eggs, nuts? Other predators do this kind of fire foraging, notably hawks.

Perhaps spreading grass fires on purpose by carrying embers to a new spot became a habit, to encourage new growth of grass to attract herds of game. Or perhaps they borrowed burning sticks to keep warm at night and only then began cooking things. There must have been a long period when fire was an optional extra, used occasionally, or used by one band and not others. It is unlikely that there was a research and development arm of this band, testing different methods of cooking, but that is what the whole habitat would have amounted to once the controlled use of fire was commonplace.

Homo erectus had discovered how to use a form of energy till now unavailable to mammals, trapped in wood and released by combustion. Human beings were thus stealing energy sources that had till now been the province of termites, fungi and bacteria. This was in effect an energy transition equivalent in its impact to the adoption of fossil fuels many millennia later.

The ultimate innovation: life itself

The beginning of life on earth was the first innovation: the first rearrangement of atoms and bytes into improbable forms that could harness energy to a purpose, which is also a good description of a car or a conference. That it happened four billion years ago, when there were no living creatures, let alone intelligent ones, and that we don't know very much about where and how it happened, does not detract from its status as an innovation. We do know that it was all about energy and improbability, both of which are crucial to innovation today. And the fact that nobody planned the origin of life is also a key lesson.

All living creatures have an idiosyncratic way of trapping energy to make it useful. Their cells pump protons across lipid membranes to create energy gradients that then fuel the synthesis of proteins that do work: they turn energy into work, just as steam engines and computers do. During every second of your life a human being pumps a billion trillion protons across membranes in the thousand trillion mitochondria that live inside the cells of the body. The failure of these proton gradients is the very definition of death. Cyanide is a poison because it blocks the proton pump. A freshly dead body is, to all intents and purposes, identical to a living one, except that on an invisible scale, its ability to keep protons the right side of membranes has suddenly ceased.

Nick Lane, of University College London, was the first to realize just how unusual this is. It seems to be an arbitrary way of making and storing energy to defy entropy locally. He guessed that it might be a clue to where and how life first emerged, a sort of fossil signature. In 2000 a new kind of alkaline, warm-water vent was found on the ocean floor in the mid-Atlantic, distinct from the acidic, black-smoker vents found elsewhere. Named the Lost City after its huge carbonate chimneys and towers, it was found to contain structures in which protons diffuse across thin, semi-conducting walls of nickel, iron and sulphur into minuscule pores. This accidental energy gradient allows, or causes, the synthesis of organic molecules, which accumulate and interact. Lane thinks life got started inside just such a pore four billion years ago. The natural proton gradients came by accident to drive the generation of molecular complexity. The origin of that energy was in the reactions between chemicals in rocks and fluids.

That the origin of life happened only once – or if it happened more than once, then the rival life form died out – is proven by the same arbitrary genetic code being found in all life forms. Thus, at the dawn of life, there was innovation by fortuitous recombination, and the result was a reduction in entropy through the harnessing of energy. Since that roughly described civilization and technology too, there is a clear sense in which human innovation is just the continuation of a process that began four billion years ago. There is no spiritual discontinuity involved here; matter has become more and more complicated, at first entirely within and then increasing without organic bodies. Some people, such as James Lovelock in his recent book *Novacene*, think that this trajectory is on the verge of a continuation that dispenses with the organic component altogether, as the robots take over and we transfer our minds to their computers.

8

Innovation's essentials

Liberty is the parent of science and of
virtue, and a nation will be great in both
in proportion as it is free.

THOMAS JEFFERSON

Innovation is gradual

The history of innovation, laid out in the stories I have told
here, reveals some surprisingly consistent patterns. Whether
it happened yesterday or two centuries ago, whether it was
high technology or low, whether it was a big device or a
tiny one, whether real or virtual, whether its impact was
disruptive or just helpful, a successful innovation usually
follows roughly the same path.

For a start, innovation is nearly always a gradual, not
a sudden thing. Eureka moments are rare, possibly non-
existent, and where they are celebrated it is with the help
of big dollops of hindsight and long stretches of preparation, not to mention multiple wrong turns along the way.
Archimedes almost certainly did not leap out of his bath,

shouting 'Heureka'; he probably invented the story afterwards to entertain people.

You can tell the story of the computer in lots of ways,
starting with Jacquard looms or starting with vacuum
tubes, starting with theory or starting with practice. But the
deeper you look, the less likely you are to find a moment
of sudden breakthrough, rather than a series of small incremental steps. There is no day when you can say: computers
did not exist the day before and did the day after, any more
than you could say that one ape-person was an ape and her
daughter was a person.

That is why it is possible to tell the stories of uncon-
scious, 'natural' innovation such as fire, stone tools and the
origin of life itself as part of a continuum with modern
technological inventions. They are essentially the same
phenomenon: evolution. In the case of the motor car, the
closer you look, the more the early versions look like older
versions of preceding technologies, like carriages, steam
engines and bicycles, reminding us that, with very few
exceptions, man-made technologies evolve from previous
man-made technologies, and are not invented from scratch.
This is a key characteristic of evolutionary systems: the
move to the 'adjacent possible' step.

Perhaps I am exaggerating. After all, there was a
moment when the Wright brothers' flier became airborne,
on 17 December 1903. Surely this was a sudden, breakthrough moment? No, far from it. Once you know the
story, nothing could be more gradual. The flight that day
lasted for a few seconds. It was barely more than a hop.
It would not have been possible without a stiff head wind
and it was preceded by a failed attempt. It came after several years of hard slog, experiment and learning, in which

very gradually all the pieces necessary for powered flight came together. Lawrence Hargreaves, an early Australian aviation experimenter, wrote in 1893 that his fellow enthusiasts must root out the idea that by 'keeping the results of their labours to themselves, a fortune will be assured them'. The genius of the Wright brothers was precisely that they realized they were in an incremental, iterative process and did not expect to build a flying machine at the first attempt. And the Kitty Hawk moment came before several more years of hard slog, tinkering and retinkering, till the Wrights knew how to keep a plane aloft for hours, how to lift off without a head wind, how to turn and how to land. The closer you examine the history of the aeroplane, the more gradual it looks. Indeed, the moment of lift-off itself is gradual, as the weight on the wheels gradually declines.

This is true of every invention and innovation I have looked at in this book so far, and of many that I have not. It is the same with the double helix, a discovery with what looks like a clear 'eureka moment' on 28 February 1953 when Jim Watson suddenly saw that the two base pairs had the same shape, Francis Crick realized that this explained the strands running in opposite directions, and they both saw how a linear digital code must lie at the heart of life. But, as Gareth Williams has written in his book on the prehistory of this work, *The Unravelling of the Double Helix*: 'this was just one episode in a long, grumbling crescendo of discovery.'

Oral rehydration therapy, the medical innovation that has saved more lives in recent decades than any other, is another good example. Some time in the 1970s in Bangladesh a number of doctors began using solutions of sugar and salt to stop children dying of diarrhoea-induced dehy-dration. Superficially, it looks like a sudden innovation. But the closer you examine the history, the more you find earlier experiments with the idea, in the Philippines in the 1960s, which themselves built upon rat experiments in the 1950s, and gradual improvements in intravenous rehydration therapy in the 1940s.

True, there followed something of an experimental breakthrough in 1967 when scientists at the Cholera Research Laboratory in Dacca (now Dhaka) in East Pakistan (now Bangladesh), led by Dr David Nalin, realized that adding glucose to a salty mixture improved the retention of sodium, but arguably they were only rediscovering the hints of studies in earlier years and testing them at scale. Similar results from Calcutta around the same time confirmed the finding. Even then the Dacca laboratory was slow to push the idea on physicians and aid workers. Some experts concluded that oral rehydration could help a little but was not a substitute for intravenous rehydration, and the conventional wisdom was that this must be accompanied by starving the gut. And when a plan to try oral rehydration in rural East Pakistan (where intravenous was not practical) was mooted in 1968, it met strong opposition from the very scientist who had first found the effect of glucose in the Philippines, Robert Philips. By the early 1970s, especially during the Bangladesh war of independence, oral rehydration therapy proved its worth as by far the best treatment for cholera and other diarrhoeas, and the innovation had arrived.

If innovation is a gradual, evolutionary process, why is it so often described in terms of revolutions, heroic breakthroughs and sudden enlightenment? Two answers: human nature and the intellectual property system. As I have shown repeatedly in this book, it is all too easy and all too tempting for whoever makes a breakthrough to magnify its importance, forget about rivals and predecessors, and ignore successors who make the breakthrough into a practical proposition.

The laurels that garland the forehead of a true 'inventor' are irresistible. But it is not just the inventor who likes to portray innovation as sudden and world-changing. So do journalists and biographers. In fact, very few people, not even the furiously disappointed rival who just failed to beat the inventor to it, have much incentive to argue that invention and innovation are gradual. As I discussed in *The Evolution of Everything*, this is, of course, a version of the 'great man' theory of history, namely that history happens because particular chiefs, priests and thieves make it happen that way. It's mostly untrue of history in general, and of the history of innovation in particular. Most people want to think they have more control over their lives than is objectively the case: the idea of decisive and discontinuous human agency is both flattering and comforting.

Nationalism exacerbates the problem. All too often, the importing of a new idea gets confused with the inventing of a new idea. Fibonacci did not invent zero, and nor did Al Khwarizmi and the other Arabs that he borrowed it from. The Indians did. Lady Mary Wortley Montagu did not invent inoculation, and probably nor did the Ottoman doctors she learned it from.

But it is the existence of patents that makes the problem of the heroic inventor worse. Again and again, I have documented in this book how innovators wrecked their lives battling to establish or defend patents on their innovations.

Samuel Morse, Guglielmo Marconi and many others tied themselves up in courts for years trying to rebut challenges to their priority. In some cases, the establishing of a patent that was too broadly drawn then deterred further innovation. This was the case with Captain Savery's patent on the use of fire to raise water, which caught Newcomen's steam engine, or Watt's patents on high-pressure steam, which slowed down improvements for some decades. I shall return in a later chapter to the point that intellectual property is now a hindrance not a help to modern innovation.

Innovation is different from invention

Charles Townes, who won the Nobel Prize for the physics behind the laser in 1964, was fond of quoting an old cartoon. It shows a beaver and a rabbit looking up at the Hoover dam: 'No, I didn't build it myself,' says the beaver. 'But it's based on an idea of mine.' All too often discoverers and inventors feel short-changed that they get too little credit or profit from a good idea, perhaps forgetting or overlooking just how much effort had to go into turning that idea or invention into a workable, affordable innovation that actually delivered benefits to people. The economist Tim Harford has argued that 'the most influential new technologies are often humble and cheap. Mere affordability often counts for more than the beguiling complexity of an organic robot.' He calls this the 'toilet-paper principle' after a simple but vital technology that we take for granted.

Fritz Haber's discovery of how to fix nitrogen from the air, using pressure and a catalyst, was a great invention. But it was Carl Bosch's years of hard experiment, overcoming problem after problem and borrowing novel ideas from other industries that eventually led to the manufacture of ammonia on a large scale and at a price that society could afford to pay. You could say the same of the Manhattan Project, or the Newcomen steam engine, but it is not only big industrial innovations that this rule applies to. Again and again in the history of innovation, it is the people who find ways to drive down the costs and simplify the product who make the biggest difference. The unexpected success of mobile telephony in the 1990s, which few saw coming, was caused not by any particular breakthrough in physics or technology, but by its sudden fall in price.

As Joseph Schumpeter put it in 1942:

> Electric lighting is no great boon to anyone who has money enough to buy a sufficient number of candles and to pay servants to attend to them. It is the cheap cloth, the cheap cotton and rayon fabric, boots, motorcars, and so on that are the typical achievements of capitalist production, and not as a rule improvements that would mean much to the rich man. Queen Elizabeth owned silk stockings. The capitalist achievement does not typically consist in providing more silk stockings for queens but in bringing them within the reach of factory girls in return for steadily decreasing amounts of effort.

Innovation is often serendipitous

The word serendipity was coined by Horace Walpole in 1754 to explain how he had tracked down a lost painting. He took it from a Persian fairy tale, 'The Three Princes of Serendip', in which, as Walpole put it in a letter, the clever princes were 'always making discoveries, by accidents and sagacity, of things which they were not in quest of'. It is a well-known attribute of innovation: accidental discovery.

Neither the founders of Yahoo! nor those of Google set out in search of search engines. The founders of Instagram were trying to make a gaming app. The founders of Twitter were trying to invent a way for people to find podcasts. At Dupont in 1938, Roy Plunkett invented Teflon entirely by accident. While trying to develop improved refrigerant fluids, he stored about 100 pounds of tetrafluoroethylene gas in cylinders at dry-ice temperatures, intending to chlorinate it. When he opened a cylinder not all of it came out: some of the chemical had polymerized and turned to a solid, white powder, polytetrafluoroethylene or PTFE. It was useless as a refrigerant, but Plunkett decided to work out what it was like. It proved to be heat-resistant and chemically inert, but also strangely friction-less, or non-stick. PTFE went on to find uses in the Manhattan Project in the 1940s, as a container for fluorine gas; as coating for non-stick pans in the 1950s; as Goretex clothing in the 1960s; and on board the Apollo missions to the moon.

Two decades later, Stephanie Kwolek developed Kevlar, also serendipitously and also at Dupont. An expert on polymers who had joined the firm in 1946, she stumbled on a new form of aromatic polyamide that could be spun into a fibre. Persuading a reluctant colleague to spin the gunky fibre into a textile, she discovered that it was stronger than steel, lighter than fibre-glass and heat-resistant. The application to bullet-proof garments only became obvious a little later. 'Some inventions,' said Kwolek, 'result from unexpected events and the ability to recognize these and use them to advantage.'

In the search for a strong and permanent glue, Spencer Silver at 3M in Minneapolis found a weak and temporary adhesive instead. This was in 1968. Nobody could think of a use for it, until five years later a colleague named Art Fry remembered it when irritated by his place-markers falling out of a hymn-book while singing in a church choir. He went back to Silver and asked to apply the glue to small sheets of paper. The only paper lying around was bright yellow. The Post-it note was born.

Or take the invention of genetic fingerprinting, a technology that has proved invaluable in the conviction of the guilty, but even more so in the exoneration of the innocent; and that has been so widely applied in paternity and immigration disputes that it is safe to say DNA unexpectedly had a far greater impact outside medicine than inside it, in the 1990s.

Alec Jeffreys, the scientist at Leicester University who made the discovery of how to use DNA to identify people and their relatives, began working on the variability of DNA in 1977, hoping to find a way of spotting gene mutations directly. In 1978 he first detected DNA variations in people, with a view to diagnosing diseases. He was still thinking in terms of medical applications. But on the morning of 10 September 1984 he realized that he had found something different. Samples from different people, including the lab's technician and her mother and father, were proving to be always different and therefore unique.

Within months the technique was being used to challenge the decisions of the immigration authorities, and to identify paternity. Then, in 1986, the Leicestershire police arrested a young man with learning difficulties, Richard Buckland. A fifteen-year-old girl had been beaten, raped and strangled in a wooded area near the village of Narborough. Buckland lived locally, seemed to know details of the crime and soon confessed under questioning to committing it. Case closed, it seemed.

The police wanted to know if Buckland had also committed a very similar crime nearly three years before and just a short distance away, in which another fifteen-year-old girl had been raped and killed. Buckland denied it. So the police asked Jeffreys, at the local university, if his new DNA fingerprinting technique could help, given that semen had been found on both bodies. Jeffreys ran a test and came back with a clear answer: the same person had committed both crimes – but it was not Buckland. The police were understandably reluctant to accept this conclusion, based on such a novel technique, but they eventually conceded that they could not convict Buckland in the light of Jeffreys' evidence and he was freed. Buckland therefore became the first person to be exonerated by DNA.

The police then asked all men of a certain age in the area to take a blood test. After eight months they had 5,511 samples. None matched the evidence from the crime scenes. A dead end. But in August 1987 a man admitted over a beer in a pub to having impersonated a work colleague when taking the test. An eavesdropper passed the news to the police. Colin Pitchfork, a 27-year-old cake decorator at a bakery, had asked his friend to take the test on his behalf, using some excuse about a previous brush with the police. The police arrested Pitchfork, who quickly confessed and whose DNA matched that found at both crime scenes.

Thus, the very first use of forensic DNA exonerated an innocent man, convicted a guilty one, and probably saved several girls' lives. Jeffreys had serendipitously set DNA on the path of making a far bigger difference in the 1990s to criminal investigation than it had made to medicine by then.

Innovation is recombinant

Every technology is a combination of other technologies; every idea a combination of other ideas. As Erik Brynjolfsson and Andrew McAfee put it: 'Google self-driving cars, Waze, Web, Facebook, Instagram are simple combinations of existing technology.' But the point is true more generally. Brian Arthur was the first to insist on this point in his 2009 book *The Nature of Technology: What It is and How It Evolves*. He argued that 'novel technologies arise by combination of existing technologies and that (therefore) existing technologies beget further technologies.' I defy the reader to find a technological (as opposed to a natural) object in his or her pocket or bag that is not a combination of technologies and of ideas. Looking at my desk as I write I see a mug, a pencil, some paper, a telephone and so on. The mug is perhaps the simplest object but even it is glazed ceramic with a printed logo and combines the ideas of baking clay, glazing, printing, adding a handle and holding tea or coffee in a receptacle.

Recombination is the principal source of variation upon which natural selection draws when innovating biologically. Sex is the means by which most recombination happens. A male presents half his genes to an embryo and so does a female. That is a form of recombination, but what happens next is even more momentous. That embryo, when it comes to make sperm and egg cells, swaps bits of the father's genome with bits of the mother's in a process known as crossing over. It shuffles the genetic deck, creating new combinations to pass on to the next generation. Sex makes evolution cumulative and allows creatures to share good ideas.

The parallel with human innovation could not be clearer. Innovation happens, as I put it a decade ago, when ideas have sex. It occurs where people meet and exchange goods, services and thoughts. This explains why innovation happens in places where trade and exchange are frequent and not in isolated or underpopulated places: California rather than North Korea, Renaissance Italy rather than Tierra del Fuego. It explains why China lost its innovative edge when it turned its back on trade under the Ming emperors. It explains the bursts of innovation that coincide with increases in trade, in Amsterdam in the 1600s or Phoenicia 3,000 years earlier.

The fact that fishing tackle in the Pacific was more diverse on islands with more trading contacts, or that Tasmanians lost out on innovation when isolated by rising sea levels, shows the intimate, mandatory connection between trade and the development of novelty. This explains too why innovation started in the first place. The burst of technology that began in dense populations exploiting rich, marine ecosystems in southern Africa more than 100,000 years ago was caused by the fact that – for whatever reason – people had begun exchanging and specializing in a way that *Homo erectus* and even Neanderthals never did. It is a really simple idea, and one that anthropologists have been slow to grasp.

Darwinians are beginning to realize that recombination is not the same as mutation and the lesson for human innovation is significant. DNA sequences change by errors in transcription, or mutations caused by things like ultraviolet light. These little mistakes, or point mutations, are the fuel of evolution. But, as the Swiss biologist Andreas Wagner has argued, such small steps cannot help organisms cross 'valleys' of disadvantage to find new 'peaks' of advantage. They are no good at climbing slopes where one must occasionally go down on the route to the summit. That is to say, every point mutation must improve the organism or it will be selected against. Wagner argues that sudden shifts of whole chunks of DNA, through crossing over, or through so-called mobile genetic elements, are necessary to allow organisms to leap across these valleys. The extreme case is hybridization. Britain alone has seven or more new species of plant that came about by hybridization in recent decades. The honeysuckle fly of North America is a new species resulting from the cross-breeding of blueberry and snowberry flies.

Wagner cites numerous studies which support the conclusion that 'recombination is much more likely to preserve life – up to a thousand times more likely – than random mutation is.' This is because whole working genes, or parts of genes, can be given new jobs, where a step-by-step change would find only worse results. Bacteria can 'catapult themselves not just hundreds of miles, but thousands of miles, through a vast genetic landscape, all courtesy of gene transfer'.

In the same way, innovation in one technology borrows whole, working parts from other technologies, rather than designing them from scratch. The inventors of the motor car did not have to invent wheels, springs or steel. If they had done, it is unlikely that they would ever have produced working devices along the way. The inventors of modern computers took the idea of vacuum tubes from the ENIAC and the idea of storable programs from the Mark 1.

Innovation involves trial and error

Most inventors find that they need to keep 'just trying' things. Tolerance of error is therefore critical. It is notable that during the early years of a new technology – the railway, for example, or the internet – far more entrepreneurs went broke than made fortunes. Humphry Davy once said 'the most important of my discoveries has been suggested to me by my failures'. Thomas Edison perfected the light bulb not by inspiration but by perspiration: he and his team tested 6,000 different materials for the filament. 'I've not failed,' he once said. 'I've just found 10,000 ways that won't work.' Henry Booth helped George Stephenson improve the *Rocket* using trial and error. Christopher Leyland helped Charles Parsons use trial and error to perfect the design of the turbine. Keith Tantlinger helped Malcom McLean get the right fit for containers on ships, by trial and error. Marconi used trial and error in his radio experiments. The Wright brothers found out by crashing that the profile of a wing should have a shallow, not a deep ratio. The pioneers of fracking stumbled on the right formula by accident and then gradually improved it by endless experiments.

An element of playfulness probably helps, too. Innovators who just like playing around are more likely to find something unexpected. Alexander Fleming said: 'I like to play with microbes.' James Watson, co-discoverer of the double helix, described his work with models as 'play'.

Andrew Geim, the inventor of graphene, said: 'a playful attitude has always been the hallmark of my research.'

A trivial example of innovation, based on trial and error: Regan Kirk of the startup Growth Tribe gives the example of Takeru Kobayashi, who in 2001 set a spectacular new record at Coney Island for hot-dog eating: he consumed fifty in ten minutes. Slim and small, Mr Kobayashi does not look like a champion hot-dog consumer, but his secret was that he worked out by systematic experimentation that he could eat the sausages faster if he separated them from the bread, and that he could then consume the buns quickly if he dunked them in water, which was not breaking the rules.

Only slightly less trivial, Dick Fosbury was a young athlete at Oregon State University who invented the 'Fosbury flop' by which he won the High Jump gold medal at the 1968 Olympics to the surprise of his more favoured competitors and the delight of the crowd. He turned over the bar on his back, head first, landing on his neck. He later described how he had used trial and error over many months to get the technique right. 'It was not based on science or analysis or thought or design. None of those things ... I never thought about how to change it, and I'm sure my coach was going crazy because it kept evolving.'

Using examples like this, Edward Wasserman of the University of Iowa has made the case that most human innovations evolve through a process that looks awfully like natural selection, rather than are created by intelligent design. Wasserman showed how the design of the violin changed gradually over time, not as a result of sudden improvements but as the result of small deviations from the norm being passed on if they worked and not if they did not. The hole in the centre of the instrument started out round, then became semi-circular, then elongated and finally f-shaped by this gradual means. Wasserman reckons this view of innovation runs into the same psychological resistance as natural selection faced in biology:

> According to this view, the many things we do and make – like violins – arise from a process of variation and selection which accords with the law of effect. Contrary to popular opinion, there is neither mystique nor romance in this process; it is as fundamental and ubiquitous as the law of natural selection. As with the law of natural selection in the evolution of organisms, there is staunch resistance to the role of the law of effect in the evolution of human inventions.

If error is a key part of innovation, then one of America's greatest advantages has come from its relatively benign attitude to business failure. Bankruptcy laws in most American states have allowed innovators to 'fail fast and fail often' as the Silicon Valley slogan has it. In some states, the 'homestead exemption' essentially allows an entrepreneur to keep his or her home if their business fails under Chapter 7 bankruptcy rules. Those states with homestead exemptions have shown more innovation than those without.

Innovation is a team sport

The myth of the lonely inventor, the solitary genius, is hard to shake. Innovation always requires collaboration and sharing, as exemplified by the fact that even the simplest object or process is beyond the capacity of any one human being to understand. In a famous essay called

'I, Pencil', Leonard Reed pointed out that a simple pencil is made by many different people, some cutting trees down, others mining graphite, others working in pencil factories, or in marketing or management, yet others growing coffee for the lumberjacks and managers to drink. Amid this vast team of collaborating people, not one person knows how to make a pencil. The knowledge is stored between heads, not inside them.

The same is true for innovation. It is always a collaborative phenomenon. (Even Australian magpies solve problems faster if they are in larger groups.) One person may make a technological breakthrough, another work out how to manufacture it and a third how to make it cheap enough to catch on. All are part of the innovation process and none of them knows how to achieve the whole innovation. Occasionally there is an inventor who is both scientifically gifted and good at business – Marconi comes to mind – but even then he or she is standing on the shoulders of others at the start, and relying on yet others later on.

The degree to which innovation is a team sport becomes ever more clear the more case histories one examines and the closer one looks at each one. The famous Green Revolution in agriculture was made possible by Norman Borlaug's astonishing diligence, determination and drive, but to tell the story as his work alone is a travesty. He got the idea of short-strawed varieties of wheat from Burton Bayles who got it from Orville Vogel who got it from Cecil Salmon who got it from Gonjirô Inazuka. Borlaug shared the hard work of selling the idea in Asia with people like Manzoor Bajwa and M. S. Swaminathan.

Terence Kealey and Martin Ricketts, in a recent paper on the Industrial Revolution, provide a long list of innovative industries that are known to have advanced by collective research and development among many actors freely sharing their ideas: the Dutch East India company's cargo ship, the *Fluyt*; Holland's windmills; Lyons' silk industry; crop rotation in England; Lancashire's cotton spinning; America's engines for steam boats; Viennese furniture; Massachusetts paper makers; a patent pool among sewing machine makers. This pattern is the rule, not the exception, and it was the flowering of societies, clubs and mechanics' institutes that gave Britain its lead in the Industrial Revolution.

Innovation is inexorable

Most inventions lead to priority disputes between competing claimants. People seem to stumble on the same idea at the same time. Kevin Kelly explores this phenomenon in his book *What Technology Wants*, finding that six different people invented or discovered the thermometer, five the electric telegraph, four decimal fractions, three the hypodermic needle, two natural selection. In 1922 William Ogburn and Dorothy Thomas at Columbia University produced a list of 148 cases of near-simultaneous invention by more than one person, including photography, the telescope and typewriters. 'It is a singular fact,' wrote Park Benjamin in 1886, 'that probably not an electrical invention of major importance has ever been made but that the honour of its origin has been claimed by more than one person.' Going further back still, it is striking that the boomerang, the blowpipe and the pyramid were all invented independently on different continents – as was agriculture.

I have documented in this book many striking examples of this phenomenon. Sure, some are evidence of collusion or conscious competition. But there is none the less a real pattern here. Simultaneous invention is more the rule than the exception. Many ideas for technology just seem to be ripe, and ready to fall from the tree. The most astonishing case is the electric light bulb, the invention of which was independently achieved by twenty-one people. There may have been a bit of snooping by some of these into the work of the others, and collaboration between them in a few cases, but mostly it is hard to find any evidence they even knew of each other's work. Likewise, there were scores of different search engines coming to the market in the 1990s. It was impossible for search engines not to be invented in the 1990s, and impossible for light bulbs not to be invented in the 1870s. They were inevitable. The state of the underlying technologies had reached the point where they would be bound to appear, no matter who was around.

The lesson this teaches throws up two paradoxes. First, the individual is strangely dispensable. If a carriage runs Swan or Edison over in their youth, or a car runs Page and Brin over, the world does not end up lacking light bulbs or search engines. Maybe things take longer, have a slightly different look and get different names. But the innovations still happen. This might seem a little harsh, but it is fairly undeniably true of every scientist and inventor who ever lived. Without Newcomen, steam engines would have surely been invented by 1730; without Darwin, Wallace did get natural selection in the 1850s; without Einstein, Hendrik Lorenz would have got relativity within a few years; without Szilard, the chain reaction and the fission bomb would have been invented in the twentieth century at some point; without Watson and Crick, Maurice Wilkins and Ray Gosling would have got the structure of DNA within months

– William Astbury and Elwyn Beighton already had got the key evidence a year earlier but did not realize it.

The paradox is that this is precisely what makes such achievements remarkable: there was a race to make them and somebody won. Individuals do not matter much in the long run, but that makes them all the more extraordinary in the short run. They emerge from among billions of rivals to find out, or make, something that any one of those billions could do. Far from being an insult, therefore, my jibe about inevitability and dispensability is actually a compliment. How incredible to be the one human being among billions who first sees the possibility of a new device, a new mechanism, a new idea. That is arguably even more miraculous than achieving something that would never be achieved by anybody else, like the *Mona Lisa* or 'Hey Jude'.

The second paradox of the inevitability of invention is that it makes innovation look predictable, yet it is not. In retrospect, it is blindingly obvious that search engines would be the biggest and most profitable fruit of the internet. But did anybody see them coming? No.

Technology is absurdly predictable in retrospect, wholly unpredictable in prospect. Thus predictions of technological change nearly always look very foolish. They either prove wildly overblown, or equally wildly underblown. Ken Olsen, the founder and chairman of Digital Equipment Corporation, was an immensely successful pioneer of 'mini-computers'. This name, in retrospect amusingly, referred to a range of machines the size of large desks, which had largely replaced computers the size of large rooms in the 1970s. So you would think that Mr Olsen would spot that computers might get smaller still and cheaper, and might eventually find uses within homes. Yet, speaking at a World

Future Society meeting in Boston in 1977, just a few years before the launch of personal computers, he reportedly said: 'there is no reason anyone would want a computer in their home.'

Likewise, in 2007, Steve Ballmer, chief executive of Microsoft, said: 'There's no chance the iPhone is going to get significant market share. No chance.' Sometimes, as the Swedish author Hjalmar Söderberg put it, you have to be an expert in order not to understand certain things.

Paul Krugman is a Nobel Prize-winning economist who in 1998 reacted to the growth of the internet, and the hype of the dotcom boom, with an article in *Red Herring* magazine entitled 'Why Most Economists' Predictions are Wrong'. He then proceeded to give a dramatic demonstration of his point by making what turned out to be a very wrong prediction himself:

> The growth of the Internet will slow drastically, as the flaw in 'Metcalfe's law' – which states that the number of potential connections in a network is proportional to the square of the number of participants – becomes apparent: most people have nothing to say to each other! By 2005 or so, it will become clear that the Internet's impact on the economy will have been no greater than the fax machine's.

It turns out that people do have a lot to say to each other. Anticipating what people want is something innovators are often good at; academics less so.

But there are also plenty of quotes from people predicting too much technological progress as well as too little. In the 1950s Isaac Asimov forecast that we would have moon colonies by the year 2000, while Robert Heinlein expected routine interplanetary travel. Others forecast supersonic rocket ships to travel around the world, human-like robots in the home and gyrocopters for all.

Innovation's hype cycle

In my view the most insightful thing ever said about forecasting innovation was a 'law' named after a Stanford University computer scientist and long-time head of the Institute for the Future by the name of Roy Amara. Amara's Law states that people tend to overestimate the impact of a new technology in the short run, but to underestimate it in the long run. Exactly when Roy Amara first had this idea is not clear. His former colleagues told me that by the middle of the 1960s he had begun making the point, and of course, in line with most innovations, this one too had its rival precursors. You can find people saying similar things all the way back to the early 1900s. It often gets credited to Arthur C. Clarke, but there is no doubt that Amara deserves most credit.

Examples abound. In the 1990s there was a period of wild excitement about the internet that then seemed to end in disappointment around the time of the dotcom bust of 2000. Where was the growth of online retail, online news and online everything that we had been promised? Well, a decade later, it was there, disrupting and destroying business models all across the retail sector, the news media, and the music and film industries, and doing so far more radically than anybody had predicted. Likewise, at the time of the sequencing of the first human genome in 2000, there were wild promises of the end of cancer and the personalization of medicine. A decade later, there was an understandable backlash: genomic knowledge seemed to have had little impact on medicine: articles asking 'whatever happened to genomic medicine?' had begun to appear. A decade after that, things are beginning to look almost as promising as the original hype.

Rodney Brooks, MIT professor turned entrepreneur, cites GPS as a classic case of the Amara hype cycle. Beginning in 1978, twenty-four satellites were launched with a goal of giving soldiers a way of locating themselves for resupply in the field. In the 1980s the program failed to deliver on its promise and was nearly cancelled several times. It began to look like a failure. Eventually, the military decided it was good enough to rely upon. It quickly spilled over into the civilian world and today GPS is so ubiquitous as to be indispensable, for hikers, map readers, farm vehicles, ships, delivery trucks, planes and pretty well everybody.

Amara's hype cycle explains a lot and it implies that, between the early disappointment and the later under-estimate there must be a moment when we get it about right; I reckon these days it is fifteen years down the line. We expect too much of an innovation in the first ten years and too little in the first twenty, but get it about right looking fifteen years ahead. The explanation for this pattern surely lies in the fact that until the invention is turned into a practical, reliable, affordable innovation, over many years, its promise remains unfulfilled.

I suspect that the Amara hype cycle can be detected today in the story of artificial intelligence, a technology whose promise has long disappointed. Thanks to graphics chips, new algorithms and lots of data, at last AI might be on the brink of not fading away. The 'AI winters' that closed down earlier bursts of excitement about machine learning may not come this time.

By contrast, I cannot help thinking that blockchain is in the early stages of the hype cycle: we are overestimating its impact in the short run. Blockchain promises to bring smart contracts that cut out middlemen, enhance trustworthiness and reduce transaction costs. But there is no way it can do so overnight in the complex ecosystem of the service economy. There is almost bound to be a burst of disappointment about what blockchain has achieved, and how many blockchain firms have failed, in around ten years' time. Yet, one day, blockchain could be huge. Facebook's Libra currency, though not a true blockchain, is undoubtedly a harbinger of things to come. Why would consumers not shift to a currency available to a third of the world population and not subject to the inflationary temptations and tax greed of politicians?

Even more is this true of self-driving cars. I keep having conversations with people who think there will be no jobs for drivers within a few short years, of trucks or taxis or limos, and that this will create so much unemployment that we need to be acting now to deal with that problem. This feels premature. The truth is that autonomous vehicles are possible but in fairly limited circumstances, and that this may not change as fast as people think, in the real world. Huge amounts of driver assistance will surely come, or are here already, so that cars can detect and avoid obstacles, cruise on motorways and freeways, parallel-park and warn the driver of delays in traffic. But in the real, messy world of crowded streets, rules and etiquette, bad weather and remote rural tracks, it is a huge jump from these kinds of increasingly smart assistance to the moment you can go to sleep at the wheel secure in the knowledge that your car will go all the way to your destination. Handing over total control of a road vehicle to a computer is a much harder problem than the equivalent in the air, for example. And then there is the need to re-engineer the entire infrastructure around roads to suit automated vehicles, not to mention the insurance market. These things take time.

I am not saying autonomous cars won't happen, just that we are likely to be underestimating the time it will take and the disappointments along the way. I am prepared to bet that ten years from now there will be stories in the media about the failed forecasts for driverless cars made in the twenty-teens; and that there will be more, not fewer, professional drivers on the planet than today. Then a decade or more after that, in the 2040s, things will indeed be changing fast. I hope to live long enough to be pleased or embarrassed by this prediction!

Innovation prefers fragmented governance

One of the peculiar features of history is that empires are bad at innovation. Though they have wealthy and educated elites, imperial regimes tend to preside over gradual declines in inventiveness, which contribute to their eventual undoing. The Egyptian, Persian, Roman, Byzantine, Han, Aztec, Inca, Hapsburg, Ming, Ottoman, Russian and British empires all bear this out. As time goes by and the central power ossifies, technology tends to stagnate, elites tend to resist novelty and funds get diverted into luxury, war or corruption, rather than enterprise. This despite empires being effectively giant 'single markets' for ideas to spread within. Italy's most fertile inventive period was in the Renaissance, when it was the small city states, run by merchants, that drove innovation: in Genoa, Florence, Venice, Luca, Siena and Milan. Fragmented polities proved better than united ones. Ancient Greece teaches the same lesson.

In the 1400s Europe rather rapidly adopted printing, a technology developed originally in China, which utterly transformed the economics, politics and religion of western Europe. The fact that Europe was politically fragmented at the time played a large role in making sure that printing caught on. Johann Gutenberg himself had to leave his home city of Mainz and move to Strasbourg to find a regime that would let him get to work. Martin Luther became a wildly successful printing entrepreneur and survived only because of the protection afforded at Wartburg by the Elector Friedrich the Wise. William Tyndale published his explosively subversive, and aesthetically beautiful, English translation of the Bible while in hiding in the Low Countries. None of these projects would have been possible in a centrally run empire.

By contrast, the Ottoman and Mughal empires managed to ban printing for more than three centuries. Istanbul, a great city of culture on the edge of Europe administering a vast empire of Christians as well as Muslims, resisted the new technology. It did so, precisely because it was the capital of an empire. In 1485 printing was banned by order of Sultan Bayezid II. In 1515 Sultan Selim I decreed that printing by Muslims was punishable by death. This was an unholy alliance: the calligraphers defending their business monopoly in cahoots with the priests defending their religious monopoly, by successfully lobbying the imperial authorities to keep printing at bay. Foreigners were eventually allowed to print books in foreign languages within the Ottoman Empire, but it was not until 1726 that a Hungarian convert to Islam, Ibrahim Muteferrika, managed to persuade the imperial authorities to allow secular (but not religious) books to be printed in Arabic. Had the lands ruled by the Sultans been fragmented into different political territories and different religions, it is impossible to believe that printing would not have happened sooner and spread faster.

In China, too, the periods of explosive innovation coincided with decentralized government, otherwise known as 'warring states'. The strong empires, most notably the Ming, effectively put a stop to innovation as well as trade and enterprise more generally. David Hume, writing in the eighteenth century, already realized this truth, that China had stalled as a source of novelty because it was unified, while Europe took off because it was divided.

America may appear an exception, but in fact it proves this rule. Its federal structure has always allowed experiment. Far from being a monolithic imperium, the states were for most of the nineteenth and twentieth centuries a laboratory of different rules, taxes, policies and habits, with entrepreneurs moving freely to whichever state most suited their project. Recently the federal government has grown stronger, and at the same time many Americans are wondering why the country is not as fleet of foot at innovation as it once was.

This fragmentation works best when it results in the creation of city states. These beasties have always been the best at incubating innovation: states dominated by a single city. For at least a thousand years, innovation has disproportionately happened in cities, and especially self-governing ones. The physicist Geoffrey West of the Santa Fe Institute made a remarkable discovery about cities. He found that cities scale according to a predictable mathematical formula called a power law. That is to say, from the population of a city he can tell you with surprising precision not just how many petrol stations, miles of electrical cable and miles of roads it will have, but how many restaurants and universities and what level of wages.

And the really interesting thing is that cities need fewer petrol stations and miles of electrical cable or road – per head of population – as they get bigger, but have disproportionately more educational institutions, more patents and higher wages – per head of population – as they get bigger. That is to say, the infrastructure scales at a sublinear rate, but the socio-economic products of a city scale at a super-linear rate. And this pattern holds throughout the world wherever Geoffrey West and his colleagues look. This fact is not true of companies. As they grow bigger, beyond a certain point they become less efficient, less manageable, less innovative, less frugal and less tolerant of eccentricity. That, says West, is why companies die all the time, but cities never do. Not even Detroit or Carthage. Sybaris was the last city to vanish altogether – in 445 BC.

Innovation increasingly means using fewer resources rather than more

The bigger cities get, the more productive and efficient they become, in terms of their use of energy to create improbability, just as the bodies of animals do: a whale burns proportionately less energy than a shrew and so lives longer, has a bigger brain and behaves in a more complicated way. London proportionately burns less energy than Bristol, has a bigger collective brain and behaves in a more complicated way. The same is true throughout the economy. Those who say that indefinite growth is impossible, or at least unsustainable, in a world of finite resources are therefore wrong, for a simple reason: growth can take place through doing more with less.

Much 'growth' is actually shrinkage. Largely unnoticed, there is a burgeoning trend today that the main engine of economic growth is not from using more resources, but from using innovation to do more with less: more food from less land and less water; more miles for less fuel; more communication for less electricity; more buildings for less steel; more transistors for less silicon; more correspondence for less paper; more socks for less money; more parties for less time worked. A few years ago Jesse Ausubel of Rockefeller University discovered the surprising and unexpected fact that the American economy has begun 'dematerializing': using not just less stuff per unit of output, but less stuff altogether. (Chris Goodall had already spotted the same to be true of Britain.) By 2015 America was using 15 per cent less steel, 32 per cent less aluminium and 40 per cent less copper than at its peaks of using these metals, even though its population was larger and its output of goods and services much larger. Its farms use 25 per cent less fertilizer and 22 per cent less water yet produce more food thanks to better targeting of fertilizer and irrigation. Its energy system generates fewer emissions (of carbon dioxide, sulphur dioxide and nitrogen oxides) per kilowatt-hour. In the ten years from 2008, America's economy grew by 15 per cent but its energy use fell by 2 per cent.

This is not because the American economy is generating fewer products: it's producing more. It is not because there is more recycling – though there is. It's because of economies and efficiencies created by innovation. Take aluminium drinks cans. When first introduced in 1959 a standard aluminium can weighed 85 grams; today it weighs 13 grams, according to Professor Vaclav Smil. This has a counterintuitive implication: those who say growth is impossible without using more resources are simply wrong. It will always be possible to raise living standards further by lowering the amount of a resource that is used to produce a given output. Growth is therefore indefinitely 'sustainable'.

The nineteenth-century economist William Stanley Jevons discovered a paradox, since named after him, whereby saving energy only leads to the use of more energy. We react to cheaper inputs by using more of them. When electricity is cheap we leave the lights on more. But Andrew McAfee, in his book *More from Less*, argues that in many sectors the economy is now exhausting the Jevons paradox and beginning to bank the savings. Thus LEDs use less than 25 per cent of the electricity that incandescent bulbs use for the same amount of light, so you would have to leave them on for more than ten times as long to end up using more power: that is unlikely to happen.

McAfee argues that dematerialization is one reason why the many pessimistic predictions of the 1970s, about the probability of running short of oil, gas, coal, copper, gold, lead, mercury, molybdenum, natural gas, oil, silver, tin, tungsten, zinc and lots of other non-renewable resources early in the current century, proved to be so spectacularly wrong: 'The image of a thinly supplied spaceship Earth hurtling through the cosmos with us on board is compelling but deeply misleading. Our planet has amply supplied us for our journey. Especially since we're slimming, swapping, optimizing and evaporating our way to dematerialization.'

9

The economics of innovation

Ideas are like rabbits. You get a couple and learn how
to handle them, and pretty soon you have a dozen.

JOHN STEINBECK

The puzzle of increasing returns

There is a curious hole at the heart of economic theory
where the word 'innovation' should be. David Warsh, in
a book on the history of economics entitled *Knowledge
and the Wealth of Nations* pointed out that Adam Smith
himself created a contradiction that he never resolved and
that in some form persists to this day. The famous 'Invisible Hand' is about the gradual emergence of equilibria in
markets, so that neither the producer nor the consumer can
improve upon the deal they have got. This implies diminishing returns: as the world settles upon the right price of a
widget, so there are no gains to be made.

By contrast, Smith's other idea, the division of labour,
implies the opposite: increasing returns. To use his own

example, in a pin factory, as workers share out the tasks and get more specialized and innovative in their work, and therefore collectively more productive, so the cost of making pins comes down and down. Both producers and consumers get more for less. The first metaphor therefore implies negative feedback, the second positive. They cannot both be right.

The economists who followed in Smith's footsteps largely forgot about increasing returns and the pin factory, focusing instead on the invisible hand. David Ricardo, Léon Walras, William Stanley Jevons, John Stuart Mill, Alfred Marshall and Maynard Keynes all more or less explicitly believed in diminishing returns. Though they lived through an era of constant innovation and accelerating prosperity, they thought the party would come to an end eventually. Mill, for instance, did not ignore technical progress, but nor did he attempt to explain it, and he assumed it would fade. Marshall had a crack at resolving this paradox. He invented the idea of 'spillovers' or positive externalities, but it was little more than a clever device to make the mathematics come out right.

Then, in 1928, an economist named Allyn Young raised the issue of Smith's contradiction, saying that the invention of new tools, new machinery, new materials and new designs involved the division of labour as well. In other words, innovation was itself a product of increased specialization, not a separate thing. He never took the idea further though. In 1942 Joseph Schumpeter argued that innovation was the main event, that increasing returns were potentially infinite: 'It is one of the safest predictions that in the calculable future we shall live in an embarras de richesse of both foodstuffs and raw materials, giving all the rein to expansion of total output that we shall know what to do with.' This was a distinctly unfashionable view even at the time, and it remains so today, though the intervening years have shown it to be true so far. Keynes, for instance, thought the Great Depression represented the arrival of diminishing returns and the need to share out less work more fairly. The trouble was that Schumpeter was not inclined to use mathematics, and economics was increasingly in thrall to the cult of the equation, so Schumpeter was largely ignored.

In 1957 Robert Solow once again raised innovation as a missing issue within economic theory. Solow argued that just 15 per cent of economic growth to date could be explained by bringing more land under the plough, bringing more workers into industry and applying more capital to investment. The residual, the 85 per cent of growth that could not be explained by these factors of production, must – obviously – be the result of innovation.

Yet even in Solow's model, innovation just arrives, like manna from heaven. It is 'external' to the model. He had no theory about why it arrived in some places and at some times rather than others. The source of this manna was later discerned by Richard Nelson and Kenneth Arrow as the government funding of research. This was something that left to itself, they argued, the private sector would not generate, because science is something that it profits nobody to create. Their argument was that a businessman will always find it easy to copy somebody else's ideas and innovations, and that the fences by which property in knowledge can be protected – patents, copyrights and secrecy – are inadequate. So the state must provide the knowledge that leads to innovation. As Professor Terence Kealey has commented, this was an ivory tower view that ignores what happens in the real world:

> The problem with the papers of Nelson and Arrow, however, was that they were theoretical, and one or two troublesome souls, on peering out of their economists' eyries, noted that in the real world there did seem to be some privately funded research happening – quite a lot of it actually.

In 1990 a young economist named Paul Romer took an interest in the problem of increasing returns and the growth of knowledge. Romer devised an answer that would eventually win him the Nobel Prize. He tried to make innovation as the source of economic growth an 'endogenous' factor in models. To put it another way he made innovation into a product, something that is an output as well as an input of economic activity. His crucial argument was that a characteristic feature of new knowledge is that it is non-rival, meaning that people can share it without using it up; but it is also partially excludable, meaning that whoever gets hold of it first can make money exploiting it, at least for a while. People can either keep new knowledge secret (as Haber and Bosch did with their iron catalysts) or patent it (as Morse did with the telegraph) or just use their 'tacit' knowledge to steal a march on their rivals in time (as most software pioneers did), and do so long enough to get a burst of partial-monopoly profits. This was a crucial distinction not made before. Knowledge is both a public good and a temporarily private one. Knowledge is expensive to produce, but can sometimes pay for itself.

Innovation is a bottom-up phenomenon

It has recently become fashionable, especially in Britain, to argue a somewhat 'creationist' view of innovation, namely that it is a product of intelligent design by government, and that government should therefore adopt an industrial policy of directed innovation. This view is championed in a 2014 book by the economist Mariana Mazzucato, *The Entrepreneurial State*, which argues that the main source of innovation has been government support of research and development with 'mission-oriented directionality'.

I find this thesis unpersuasive, and detailed critiques of it, especially by Alberto Mingardi and Terence Kealey, persuasive. Here is why. As this book has documented, innovation is not a new phenomenon. It was responsible for the dramatic improvements in human living standards that emerged in the nineteenth century and earlier. Yet the technologies and ideas behind this 'great enrichment' owed little or nothing to government. Throughout the nineteenth century, as Britain and Europe developed new railways, steel, electricity, textiles and many other technologies, government played almost no role at all except as a belated regulator, standards creator or customer. Mazzucato specifically cites railways as an example of public innovation, but the British and global railway boom of the 1840s was an entirely private-sector phenomenon, notoriously so: fortunes were made and lost in bubbles and crashes. At the time almost the entire state budget of Britain was spent on defence and servicing the debt incurred in war, with effectively none on innovation, let alone in a mission-oriented way. Yet railways transformed people's lives. As William Makepeace Thackeray wrote:

Bless railroads everywhere
And the world's advance
Bless every railroad share
In Italy, Ireland, France;
For never a beggar need now despair,
And every rogue has a chance.

The economic historian Joel Mokyr argues that 'any policy objective aimed deliberately at promoting long-run economic growth would be hard to document in Britain before and during the Industrial Revolution'. It would be strange to argue that innovation could happen without state direction in the nineteenth century, but only with it in the twentieth.

The same is true of America, which became the most advanced and innovative country in the world in the early decades of the twentieth century without significant public subsidy for research and development of any kind before 1940. The few exceptions tend to confirm the rule: for example, the government heavily subsidized Samuel Langley's spectacular failure to make a powered plane, while wholly neglecting the Wright brothers' spectacular success, even after they had proved their point.

In a parallel case, some years later, in 1924 Britain's new Labour government decided it needed to design and build airships that could cross oceans, a feat that was at that time considered beyond the reach of conventional heavier-than-air planes carrying passengers. Experts urged the government to give the contracts to private firms, but, being socialist, they resisted and eventually decided to run a controlled experiment of two different approaches: a privately funded R100 built by Vickers, and a publicly funded

R101 built by the government. Mission-oriented innovation indeed. The result was unambiguous. The R100 was lighter, faster and ready sooner. It flew to Canada and back in the summer of 1930 without mishap. The R101 was late, extravagant, over-engineered, underpowered, plagued by gas leaks and hastily redesigned at the last minute to give it more lift. On its maiden flight to India, in October 1930, carrying the air minister, it got as far as northern France and crashed, killing forty-eight of the fifty-four people on board, including the minister. A plaque records to this day where the forty-eight bodies lay in state in Westminster Hall. Neville Shute, later a novelist, who was an engineer in the R100 programme, was scathing in his book *The Slide Rule* about the failures of the nationalized project: 'I was thirty-one years old at the time of the R101 disaster, and my first close contact with senior civil servants and politicians at work was in the field of airships, where I watched them produce disaster.'

In the second half of the twentieth century, the state did become a sponsor of innovation on a large scale, but that is hardly surprising given that it went from spending 10 per cent of national income to 40 per cent in almost all Western countries. As Mingardi put it: 'With such extraordinary growth, it is improbable that public spending wouldn't end up in the neighborhood of innovation-producing business at one point or another.' So it is not a matter of whether the state has caused some innovation. The question is whether it is better at doing so than other actors, and whether it does so in a directed fashion. I have shown in this book that many of the technologies that the nation state gave a push to during the Second World War – computing, antibiotics, radar, even nuclear fission – originated in peacetime and would probably have developed just as fast if war had not broken out, maybe faster, with the probable exception of fission.

Moreover, Mazzucato's examples of government-funded innovation are mostly cases of 'spillover', rather than direction. Nobody has claimed that government set out deliberately to create a global internet when it funded the Defense Advanced Research Projects Agency's computer networking. Indeed, the internet only took off when it eventually escaped the clutches of the Defense Department and was embraced by universities and businesses. In addition, although some of the key technologies of the internet, including packet switching, originated in public institutions, others came from the private sector. The TCP/IP protocol came from Cisco, while glass fibre came from Corning.

Mazzucato points out that the technology behind the touch-screen that is crucial to the modern smartphone was invented by a PhD project in a public university, that of Wayne Westerman at the University of Delaware. But that was a very small part of the development of this idea into a useful innovation, the rest of which happened in the private sector, and it is the very opposite of directional funding: a National Science Foundation grant for research on a topic chosen by the university and the student. We must beware of down-playing the development of technologies after they are first invented, a huge part of innovation, lest we credit a beaver with the Hoover dam.

Mazzucato also cites the Small Business Innovation Research programme started by Ronald Reagan as an example of government-funded innovation in the private sector. Yet Mingardi points out that this is the very opposite of directed innovation. The programme simply requires all government agencies with an R&D budget over $100m to spend 2.8 per cent of their budget to promote innovation by small- and medium-sized businesses.

The Japanese government is sometimes cited as an example of an entrepreneurial state, which between 1950 and 1990 drove huge economic success on the back of directed innovation. This is also a myth. Until 1991, according to Terence Kealey, the Japanese government 'was funding less than 20 percent of its R&D and, remarkably, less than half of its country's academic science – an extraordinary exception to the average OECD government, which was funding around 50 percent of its R&D and 85 percent of its country's academic science'. The Japanese miracle was a function of private corporations backed by a vast ecosystem of small enterprises.

By contrast the Soviet Union was a very clear case of an entrepreneurial state, funding a great deal of research centrally, allowing virtually no private enterprise, and the result was a dismal lack of innovation in transport, food, health or any consumer sector, but lots of advances in military hardware.

In 2003 the OECD published an inconvenient paper for the entrepreneurial-state argument. Called 'Sources of Economic Growth in OECD Countries', it systematically reviewed the factors contributing to growth between 1971 and 1998 and found that whereas the quantity of privately funded research and development did affect the rate of economic growth, the quantity of publicly funded R&D did not. This was a shocking finding, probably best explained by the phenomenon of 'crowding out': government spending on research diverts the energy of researchers into its priorities, which might not coincide with industry's or the consumer's (spectacularly so, in the case of the Soviet Union). In the words of Walter Park of the American University: 'the direct effect of public research is weakly negative, as might be the case if public research spending has crowding-out effects which adversely affect private output growth.' Mazzucato acknowledges this phenomenon of crowding out when she writes that the 'top pharmaceutical companies are spending decreasing amounts of funds on R&D at the same time that the state is spending more'.

It is not, of course, impossible that governments can aim for, create and perfect an innovation of huge significance without much private input. Nuclear weapons might be one example, moon shots another, though hardly ones with any consumer value, and both in practice used a lot of private-sector contractors. It is just that it does not happen very often, and that far more often inventions and discoveries emerge by serendipity and the exchange of ideas, and are pushed, pulled, moulded, transformed and brought to life by people acting as individuals, firms, markets and, yes, sometimes public servants. Trying to pretend that government is the main actor in this process, let alone one with directed intentionality, is an essentially creationist approach to an essentially evolutionary phenomenon. After all, even the chain reaction of nuclear fission built upon a key insight that supposedly occurred to an unemployed refugee waiting at a traffic light on Southampton Row in London on 12 September 1933: Leo Szilard. And in many cases government actively gets in the way of technology. See the example of mobile telephones, explored in more depth in Chapter 11, when America's government regulations blocked the development of cellular phones for decades, and Europe's explicit adoption of an industrial policy for 2G networks trapped the continent in a standard that was soon overtaken by America.

There is another problem with the thesis that government is the source of innovation. The argument nearly always turns on the things that governments supposedly invent and then spin out into the private sector. But, if this were the case, would not governments apply them first within government itself? There is nothing quite so lacking in innovation as the practices and premises of government. Garry Runciman once argued that if Daniel Defoe was revived, three centuries after publishing his *Tour thro' the whole island of Great Britain* in 1727, after he had got over his astonishment at our cars, planes, skyscrapers, jeans, toilets, smartphones and working women, about the only things he would find familiar would be Parliament and the monarchy. The rest would be dazzlingly different. Parliament is a sociological coelacanth, a living fossil little changed since the political Paleozoic. This is not altogether a bad thing – we make it hard to disrupt traditional ways of making laws for good reason – but it hardly speaks of a society that breeds innovation from government outwards.

I repeat that none of this should be taken to imply that government is incapable of stimulating innovation, or that everything it does is better done by other actors. The advertising executive Rory Sutherland cites the example of video-conferencing, as a technology that the British state might usefully push, through the roll-out of high-speed broadband, because a traffic-congested, English-speaking nation in its time zone could disproportionately benefit from it, and because the government could negotiate a collective price for the work, unleashing the network effect by which such things become really useful only when lots of other people use them: the free-rider opportunity, rather than the free-rider problem. Such opportunities undoubtedly exist. But it is a myth to think that government uniquely and deliberately caused most recent innovation.

Innovation is the mother of science as often as it is the daughter

There is a widely held view among politicians, journalists and the public that science leads to technology, which leads to innovation. This 'linear model' holds sway amongst almost all policy makers and is used to justify public spending on science, as the ultimate fuel of innovation. While this can sometimes happen, it is just as often the case that invention is the parent of science: techniques and processes are developed that work, but the understanding of them comes later. Steam engines led to the understanding of thermodynamics, not the other way round. Powered flight preceded almost all aerodynamics. Animal and plant breeding preceded genetics. Pigeon fancying laid the groundwork for Darwin's understanding of natural selection. Metalworking helped give birth to chemistry. None of the pioneers of vaccination had the foggiest idea how or why it worked. Understanding of the mode of action of antibiotics came long after they were in practical use.

In 1776 Adam Smith was well aware of the primacy of practice. He saw innovation as deriving from the tinkering of 'common workmen' and 'the ingenuity of the makers of the machines'. These were far more important than academic research, for although 'some improvements in machinery have been made by those called philosophers', philosophy had got more from industry than vice versa: 'The improvements which, in modern times, have been made in several different parts of philosophy, have not, the greater part of them, been made in universities.'

This surely changed in recent decades. Yet in the 1990s Edwin Mansfield, surveying companies to identify the sources of their innovations, found that nearly all originated in house or within the industry. In the case of new processes (just as important as new products) just 2 per cent of innovations originated in academia. Universities contributed little to ideas about industrial organization, for instance. And when science does spawn new industries, there is often a reciprocal effect: science helps technology along, which in turn helps science along. More recent work finds that about 20 per cent of patents cite academic scientific work, about 65 per cent of patents have some connection to scientific research, and the scientific fields that do more basic research end up with more approved drugs and patents. But this still does not prove a linear connection as opposed to a two-way relationship.

Thus the discovery of the structure of DNA in 1953 is a good example of pure science with huge later practical implications, in superficial accordance with the linear model, but it owed a lot to the development of X-ray crystallography on the structure of biological molecules. That work was begun and funded partly by the textile industry to try to understand more about the properties of wool. This was what took William Astbury to Leeds University, whose chemistry department was 'essentially a finishing school for those going into the textile industry', as Gareth Williams put it. Astbury's post was funded by the Worshipful Company of Clothworkers. Astbury, together with Florence Bell and Elwyn Beighton, was the first to start to understand the structure of proteins and DNA using X-rays. Beighton actually beat Raymond Gosling and Rosalind Franklin by a year to taking a photograph that revealed the structure – if only he had realized it. His photograph, and its significance, was overlooked by Astbury in one of the saddest near-misses of all time.

Likewise, in the twenty-first century, as I have documented in Chapter 4, the work that led to the invention of CRISPR gene editing was driven partly by a desire to solve practical problems in the yogurt industry. My point is that we make a mistake if we insist that science is always upstream of technology. Quite often scientific understanding comes from an attempt to explain and improve a technical innovation.

The 'linear model' is actually a bit of a straw man. Though politicians are wedded to it, no economist or scientist really believes in it. As David Edgerton, a historian of technology, has argued, even those who were supposed to have invented it did not espouse it. For example, Vannevar Bush, wartime scientific adviser to the United States government, wrote a book in 1945 called *Science: The Endless Frontier*, which is supposed to be the bible of the linear model. It is true that Bush wrote: 'There must be a stream of new scientific knowledge to turn the wheels of private and public enterprise' and 'today, it is truer than ever that basic research is the pacemaker of technological progress.' But he was in fact arguing for state support for basic science in its own right, on the grounds that the growth in such research had not kept up with the growth in applied research in both government and industry. Bush did not claim that academic science is the main source of innovation, so much as the main source of new knowledge. America at the time, unlike most European countries, barely supported basic science from the federal government. Britain, too, was slow to provide public support for science compared with France and Germany, but as Kealey observes: 'the continent supposed markets failed in science, the UK supposed they did not, and the industrial revolution was British, not French or German.'

Britain's wartime science chief, Sir Henry Tizard, wrote after the war that 'it is not the general expansion of research in this country that is of first importance for the restoration of its industrial health, and certainly not the expansion of government research remote from the everyday problems of industry. What is of first importance is to apply what is already known' – thus becoming the first in a long line of Britons to lament the country's modern inability to translate research strength into competitive innovation success. In 1958 an influential book called *The Sources of Invention* by the economist John Jewkes argued against the idea that science was the source of technology and warned governments against investing in pure science in the hope of stimulating economic growth. Edgerton, writing in 2004, was blunt: 'My claim is then, that the "linear model" did not exist in even the earliest generations of academic work on innovation.'

Yet there is no doubt that in recent years there has been a growing tendency among politicians to adopt the notion that science is the mother of invention, and that this is the main justification for funding science. This seems to me a pity, not just because it misreads history, but because it devalues science. To reject the linear model is most definitely not an attack on the funding of science, let alone on science itself. Science is the greatest fruit of human achieve-ment, bar none, and deserves rich and enthusiastic support in any civilized society, but as a worthwhile goal in its own right, not just as a way to encourage innovation. Science should be seen as the fruit rather than the seed. In 1969 the physicist Robert Wilson, testifying to the US Senate about funding for a particle accelerator, was asked if it would contribute to national defence. He replied: 'it has nothing to do directly with defending our country except to help make it worth defending'. In recent years there has been a tendency to demand of academic scientists that they jus-tify their financial support from the taxpayer by showing that their work generates applied spin-offs. Frankly, to ask Stephen Hawking to show that research on black holes led to industrial activity, or Francis Crick to justify research on DNA on similar grounds, would have been like asking William Shakespeare or Tom Stoppard to show that their plays contributed to economic growth. They might do, but that is hardly their main point.

Innovation cannot be forced upon unwilling consumers

Innovation is not necessarily a good thing. It can be harmful if it results in toxic or dangerous products. Fritz Haber not only invented synthetic fertilizer; he also invented poison gas for use in the trenches. Innovation can also be useless for ordinary people. Manned space travel has so far proved useful as exploration and entertainment – the moon landing and the films based on the Apollo missions are fine add-itions to human culture – but not as a source of significant economic benefit. Sure, there may have been some spin-offs in terms of understanding how to develop certain technologies (though the non-stick frying pan as one of them is a myth: if anything the reverse is true and Teflon proved vital to Apollo), but it is hard to be sure such spin-offs would not have occurred anyway in other ventures if similarly huge budgets had been spent on them. For example the Jet Propulsion Laboratory boasts that camera phones, CAT scans, LEDs, athletic shoes, foil blankets, home insulation, wireless headsets and freeze-dried food are all examples of things we would not have without space travel, because at some point in their development somebody in the space programme contributed to their development. That is a non sequitur and a very doubtful claim.

Again, this is not to carp at the space programme: I remain deeply grateful to Neil Armstrong and the American taxpayers for generously expanding my universe of knowledge and imagination and giving me a thrilling moment when I was eleven years old. But manned space exploration does not really meet the test of an innovation that is ever likely to pay for itself. It is more like art in that sense. To contribute to human welfare, and therefore catch on without subsidy, an innovation must meet two tests: it must be useful to individuals, and it must save time, energy or money in the accomplishment of some task. Something that costs more to buy than an existing device, but offers no extra benefits, will not thrive, however ingenious. Manufacturing in space may never pass this test, because of its cost.

Innovation increases interdependence

What does innovation add to people's lives? The grand theme of human history, I have argued, is the increasing specialization of production combined with the increasing diversification of consumption. We have gradually got narrower and narrower in what we produce – we call it a job – in order to get more and more varied in what we consume. Compared with a subsistence farmer, most modern people have a less varied job but a much more varied life. This is in contrast to other animals that only consume what they produce. Throughout history, economic downturns – from the fall of the Roman Empire to the Great Depression – have been characterized by retreats towards greater self-sufficiency. By contrast economic advances – from the invention of farming to the mobile internet – have been accompanied by the increased interdependence and co-operation that comes from selling a specialized service and buying everything else: working for each other. Innovation has enabled both the narrowing of 'work' and the broadening of everything else. Based on this, it is reasonable to expect that something new that increases the specialization of production or increases the diversity of consumption will catch on, something that returns us towards self-sufficiency will not catch on.

Hang on, I hear you say, but has the internet not done the opposite? Instead of having a travel agent book your flights, you now do it yourself. Instead of having a typist take down your dictation, you now use your own keyboard. Well, when you think about it, this is the perspective of the upper middle class, who had access to things like travel agents and secretaries. The internet brought travel agents to everybody in the form of websites. It brought 'secretaries' in the form of word-processing programs complete with spellchecks, formatting and graphics.

This is why today, unlike a century ago, a plutocrat is hard to spot in a crowd, as the economist Don Boudreaux has pointed out. Next time you are in a restaurant, look at the person at the next table. Is he or she a billionaire? Unlikely, but how could you tell? The bodyguards, the limo parked outside, the private-jet logo on the jacket? Big deal: these are all luxuries. What about necessities? Does he or she have better teeth? Longer legs? Greater girth? Better clothes? Fewer holes in her trousers (more likely the reverse today!)? All of these would have been true two centuries ago. Not today. She uses a similar smartphone, the same internet, the same sort of toilet, the same supermarkets. For most people today in Western countries, much of the inequality that exists – though not all – is about luxuries, rather than necessities; at least, this is more true than it was in the past, when poor people often starved to death or died of cold and lacked access to simple things like light. This is why rich people talk a great deal about things like wine and property, two forms of luxury where the sky is the limit in terms of differentiation, and not about trousers and books, which almost everybody can afford. Innovation did that by raising the productivity of work and therefore the living standards of all.

Innovation does not create unemployment

The fear that innovation destroys jobs has a long history, dating back to General Ludd and Captain Swing in the early 1800s. In 1812 the Luddites went about smashing stocking frames in protest at the introduction of new machinery into the textile industry, taking their inspiration and their name from a probably apocryphal story of one Ned Ludd who had supposedly done the same back in 1779. In 1830, in a protest at conditions in the farming industry, rioting labourers began burning hay ricks and smashing threshing machines, under a mythical leader named Captain Swing. This too was a protest at the effect of machines on livelihoods. The economist David Ricardo became 'convinced that the substitution of machinery for human labour is often very injurious to the interests of the class of labourers'. Yet far from leading to widespread rural misery, the advent of machinery saw farm workers' wages generally increase and the surplus manpower rapidly taken up by jobs in the towns to supply goods to people with more disposable income.

The idea of technology causing unemployment did not go away. John Maynard Keynes worried in 1930 that 'the increase of technical efficiency has been taking place faster than we can deal with the problem of labour absorption'. In 1960 a recession caused a rise in unemployment in America, and *Time* magazine reported that 'many a labor expert tends to put much of the blame on automation' and it would get worse: 'what worries many job experts more is that automation may prevent the economy from creating enough new jobs.' By 1964 President Lyndon Johnson had created a National Commission on Technology, Automation and Economic Progress to investigate whether innovation would destroy work. By the time it reported in February 1966, American unemployment had fallen back to just 3.8 per cent. None the less, the commission recommended drastic action to share out the remaining work fairly, including a guaranteed minimum income and the government as employer of last resort, because of the 'potentially unlimited output by systems of machines which will require little co-operation from human beings'.

In short, the idea that innovation destroys jobs comes around in every generation. So far it has proved wrong. Over

the past two centuries productivity in agriculture dramatically increased, but farm workers moved to cities and got
jobs in manufacturing. Then productivity in manufacturing
rocketed upwards, freeing huge numbers of people to work
in services, yet still there was no sign of mass unemployment. Candles were replaced by electric lights, but wick
trimmers found other work. Millions of women joined the
paid workforce, at least partly as a result of innovation in
washing machines and vacuum cleaners, which freed them
from much household drudgery, yet employment rates went
up not down. In 2011 President Obama used the bank teller
as an example of a job that has disappeared because of cash
machines. He was wrong: there are more tellers employed
today than before cash machines were introduced, and their
jobs are more interesting than just counting out cash. On
the day I write this, the percentage of working-age people
in paid employment in Britain has just reached a record
high of 76.1 per cent.

Today, it is innovation in artificial intelligence that supposedly threatens to put many people out of work. This
time is different, say many. This time, it is the cognitive
skills of the machines, rather than their brute force, that
rival those of people, leaving workers nowhere to go. By
this you mean, I sometimes reply to academics or politicians who make this case, that it is intellectuals like you
– and lawyers and doctors – who are now threatened, not
just farm labourers, housewives or factory workers. There
is a degree of special pleading going on here.

One especially influential study, by Carl Frey and
Michael Osborne, published in 2013, came to the conclusion that 47 per cent of all jobs in America are at 'high
risk' of automation within a 'decade or two'. However, the

OECD re-examined the issue, used a more appropriate database and concluded that a much less frightening 9 per cent of jobs were at risk of disappearing because of automation, and even these would be accompanied by expanding employment in other occupations. But the scary scenarios are often popular with politicians and journalists. As the economist J. R. Shackleton observes: 'technophobic panic is already tempting policy-makers to consider untested policies that are often being pushed by political activists for reasons which have little to do with a threat to existing jobs.' A recent survey found that 82 per cent of Americans think that over the next thirty years robots and computers will 'probably or definitely do most of the work done by humans' but that only 37 per cent think they will do 'the type of work I do': a big contradiction there.

The truth is, there is nothing unusually fast, or sweeping or threatening about innovation today, as it affects work. After all, as Adam Smith pointed out, the purpose of production is consumption; the purpose of work is to earn enough to get the things you want. Enhanced productivity means enhanced ability to acquire the goods and services you need, and therefore enhanced demand for the work of those that supply them. It is only the high productivity, and therefore high spending power, of the average modern worker that keeps restaurant chefs and pet vets and software experts and personal trainers and homeopaths in business.

Innovation also creates wholly new kinds of jobs. Most of the jobs that people do today would sound utterly baffling to a Victorian. What is software, or a call centre, or a flight attendant? Innovation frees people to do the things they really value. Instead of digging and weeding your vegetable patch to avoid starvation, you can choose to go out to work and buy veg from a shop. That is made possible by high productivity at your work. Walter Isaacson concludes that: 'Advances in science when put to practical use mean more jobs, higher wages, shorter hours, more abundant crops, more leisure for recreation, for study, for learning how to live without the deadening drudgery which has been the burden of the common man for past ages.'

Moreover, many people fail to notice that automation does create extra leisure, and that instead of forcing that leisure upon the unemployed, we share it out fairly equitably. Here is a fascinating fact. In 1900, when the average lifespan in the United States was forty-seven, when people started work at fourteen, worked sixty-hour weeks and had no possibility of retirement, the percentage of his lifetime that an average man would spend at work was about 25 per cent: the rest was spent sleeping, at home or as a child. Today that figure is about 10 per cent, because the average person lives to about eighty, spends about half his or her life in education and retirement, spends only a third of each day (8/24) and five-sevenths of each week at work. A half times a third times five-sevenths is just under 12 per cent. Take off a few weeks' vacation, a little sick leave and the usual holidays like Christmas and you are left with 10 per cent. And that's counting the lunch hour as work. So, yes, on the whole society has decided to use the greater productivity provided by innovation to give everybody a lot more leisure. When John Maynard Keynes forecast that Westerners would only have to work fifteen hours a week as a result of automation, or Herman Kahn forecast that we would go down to four-day weeks with thirteen weeks of vacation, they were not as wildly out as you might imagine.

As Tim Worstall observes: 'Do you or does anyone else have absolutely everything you can even dream of desiring which requires human work to deliver to you? No? Still short of a back rub, the peeled grape? Then there's still a task or two for humans to do.' Imagine if robots could do literally everything you conceivably wanted done – including back rubs and grape peeling – and could do them all so cheaply you've no need to go out to work to earn anything. What exactly is the problem? You can summon up whatever goods or services you want at zero cost. So you don't need to earn a living, because a living is free. This is not going to happen, of course, not least because there are always going to be things that you can think of doing that the robot cannot (do you really want a robot to play tennis for you?), and robots are never going to be wholly without cost, if only because they need energy, but it's a useful thought experiment. Work is not an end in itself.

Big companies are bad at innovation

Innovation often comes from outsiders. This is true of individuals as well as organizations. John Harrison was just a Yorkshire clockmaker, and when he solved the problem of establishing longitude by building robust chronometers for ships, the establishment refused to take him seriously for a long time, because he was not a scientific grandee and his solution did not involve advanced astronomy. From Thomas Newcomen to Steve Jobs, again and again the great innovators have come from obscure origins, in unfashionable provinces and without good contacts or education.

At the other end of the scale, big organizations are also frequently undone by more innovative startups. IBM was blindsided by Microsoft and Microsoft by Google and Apple. Kodak did not develop digital photography, despite having a strong position in film. Instead it watched in paralysed horror as its entire business model was disrupted to extinction by interlopers from the electronics industry. It filed for bankruptcy in 2012. Actually, this version of events is not quite true. Kodak did invent digital photography but had too big a vested interest in hoping it would go away, rather than exploring it. In 1975 a young Kodak researcher named Steven Sasson built a bulky camera that recorded a fuzzy electronic image on to a cassette tape so it could be displayed on a television screen. He tried to interest executives in his invention, but they protested that it was expensive, impractical and poor quality. Sasson told *The New York Times*: 'Print had been with us for over 100 years, no one was complaining about prints, they were very inexpensive, and so why would anyone want to look at their picture on a television set?'

Big companies are bad at innovating, because they are too bureaucratic, have too big a vested interest in the status quo and stop paying attention to the interests, actual and potential, of their customers. Thus for innovation to flourish it is vital to have an economy that encourages or at least allows outsiders, challengers and disruptors to get a foothold. This means openness to competition, which historically is a surprisingly rare feature of most societies. Monarchs were addicted throughout history to granting monopolies, whether to trading companies, craft guilds or state enterprises.

The one thing that does make a big company innovate is competition. Supermarkets, led by companies such as Walmart, Tesco and Aldi, brought their customers a constant stream of innovations during recent decades: bar-codes, scanners, truck-to-truck loading docks, pre-washed salad, ready meals, own-brand products, loyalty cards and more. There is zero doubt that if these firms had been national monopolies, innovation would have been slower or non-existent. And much of the innovation in the retail sector comes from outside the sector: firms are alert to new technologies that they can exploit.

Some big companies recognized a few years ago that they could not rely on in-house research and development to bring them the innovations they needed to compete. Procter and Gamble is a good example. As two executives explained in 2006:

> By 2000, it was clear to us that our invent-it-ourselves model was not capable of sustaining high levels of top-line growth. The explosion of new technologies was putting ever more pressure on our innovation budgets. Our R&D productivity had leveled off, and our innovation success rate – the percentage of new products that met financial objectives – had stagnated at about 35%.

The chief executive, A. G. Lafley, set out to change P&G's culture, by obtaining half of all innovations from outside the firm. This 'open innovation' strategy had the desired effect, with P&G reviving the rate at which it launched successful new products.

The ultimate form of open innovation is open-source software. It was once an eccentric, Bohemian branch of the industry, populated by communitarian dreamers who wanted the world to have no frontiers and no property. Richard Stallman's Free Software Foundation in the 1980s began a rebellion against the proprietary software of big firms and bet on the idea that users could contribute to innovation. He developed GNU (standing for Gnu's Not Unix) to challenge the position of the Unix operating system. In 1991 Linus Torvalds invented the open-source Linux operating system, incorporating features of GNU, which gradually took over much of the computing world, gaining total dominance of the supercomputer market and more recently colonizing the mobile market through Google's Android devices. In 2018 IBM announced it would pay about $30bn for an open-source software firm, Redhat. Amazon's domination of the cloud, through Amazon Web Services, is based entirely on open-source software. Thus the software world is increasingly a place of open, free sharing of innovations, an unfenced prairie. Far from deterring innovation, the effect seems to have been to encourage it. The Linux Foundation now hosts thousands of open-source projects to 'harness the power of open source development to fuel innovation at unmatched speed and scale'.

Setting innovation free

The ultimate open-source innovation is that done by consumers themselves. Eric von Hippel of Massachusetts Institute of Technology argues that free innovation by the consumer is a neglected sector of the economy and the assumption that innovation is driven by producer innovation is misleading. He calculates that tens of millions of consumers spend tens of billions of dollars every year developing or modifying products for their own use. Most do so during their free time and share them freely with others. He gives the example of Nightscout, a technology for monitoring the sugar levels in diabetics via the internet. Nightscout is the brainchild of several parents of diabetic children. A company called Dexcom developed the sensors that record blood sugar levels from skin patches and display them on a pager. In 2013 a supermarket software engineer in Livonia, New York, John Costik, worried about knowing his young son's glucose level while he was at school, hacked the device so he could see his son's data on the web, then shared his source code with others via social media. One of these was an engineer in northern California, Lane Desborough, who also had a son with diabetes and designed a home-display system, coining the name Nightscout. That in turn was seen by a molecular biologist in southern California named Jason Adams, also with a diabetic son, who developed and set up a Facebook group for parents using the Nightscout system. Only concern about regulations and intellectual property held them back from publishing open-source code earlier than they did. Nobody here was after a profit. The latest incarnation of this trend is an artificial pancreas for diabetics with open-source software, designed and evolved by patients themselves.

The opportunities for free innovation are growing as computerized design tools and lowered communication costs allow people to do at home what once required a corporate laboratory to achieve. Unhindered by doubts about whether an innovation can be profitable, free innovators can explore ideas that companies will not touch. But since they are not after profits, they do not necessarily try hard to tell people about their inventions, so diffusion can be slow. Von Hippel says that the growth of free innovation has not gone unnoticed by firms, which now 'mine' the ideas of their consumers. Makers of surfboards, for example, use surfer modifications as clues to what they should do to their designs.

Free innovators rarely seek patents or copyrights, which means they are willing to share ideas freely. Von Hippel's colleague, Andrew Torrance, argues that under the common law and the US constitution, individuals have fundamental legal rights to engage in free innovation, to use their innovations and to disclose and discuss them, under the right to free speech. This has not stopped politicians throwing obstacles in their way. The Digital Millennium Copyright Act (DCMA) of 1998, intended to clamp down on free copying or 'piracy' has caused severe collateral damage to free innovators' abilities to innovate by hacking software that was legally bought, argue Torrance and von Hippel. The DMCA effectively reduced the room for 'fair use' exceptions to copyright infringement. Those who drafted the bill were apparently unaware, Torrance argues, of the existence of free innovation, let alone the damage the legislation might do to it.

Intriguingly, Torrance and his zoologist colleague Lydia Hopper have pointed out that free innovation by users for their own benefit is the only kind indulged in by non-human animals. That is to say, there are no such things in the non-human world as producers and consumers.

10

Fakes, frauds, fads and failures

> We need big failures in order to move the
> needle. If we don't, we're not swinging enough.
> You really should be swinging hard, and you
> will fail, but that's okay.
>
> JEFF BEZOS

Fake bomb detectors

Innovation has wrought such miracles that it is little
wonder it sometimes attracts fakers, frauds, faddists and
failures, people who promote particular innovations either
in the knowledge they are not going to work, or inno-
cently hoping that they will and not succeeding. Recall
that Enron, an energy company that tried to turn itself into
an online energy-trading platform and then into a general
commodity-trading business, was named 'America's Most
Innovative Company' by *Fortune* magazine in six consecutive years between 1996 and 2001. By the end of 2001 it was
bankrupt as the illegal accounting that had hidden its losses

off balance sheet unravelled. The loss to shareholders was over $74bn. Enron's executives kept up a constant stream of exaggerated promises about the dividends of innovation, right to the end.

If you are going to start a fake-innovation scam, Wade Quattlebaum sounds like an appropriately implausible name but in fact it is his real name. In the 1990s this American car dealer and occasional treasure hunter started marketing an innovation called the Quadro Tracker Positive Molecular Locator. It was a slicker version of a device known as the Gopher, which supposedly helped people find lost golf balls. His version was also useful, Quattlebaum said, for locating drugs and explosives. It had a free-swinging antenna, like a dowsing rod, attached to a pistol grip linked to a box worn on the belt.

A gullible person could just about convince himself that the swinging antennae were moving under the influence of some signal rather than because of the movements of the hand – the same self-deception, or 'ideomotor response', that lies behind Ouija boards. Even so, it is hard to imagine anybody fell for such a transparent scam as the claim that this device could detect golf balls, let alone explosives and drugs too, but some people did. Quattlebaum signed up several salesmen to sell the devices to schools for detecting drugs before the FBI intervened and a judge banned the Quadro as a fake. In 1997 Quattlebaum and three associates were prosecuted on three counts of mail fraud and one count of conspiracy to commit mail fraud. They were found not guilty on a technicality.

A trivial episode, but it was soon to get more serious. The company's secretary, Malcolm Stig Roe, had jumped bail and moved to Britain. There he marketed a new version of the same thing and sold it to police forces. A retired British policeman named Jim McCormick signed up as a distributor, then decided to make a bigger, better and more expensive version himself. By 2006 McCormick, not without difficulty, had managed to persuade a factory to manufacture his 'ADE 650' devices and called himself Advanced Tactical Security & Communications Ltd. He promptly sold five for $10,000 each to the Lebanese army, which ordered eighty more, and other governments soon followed suit. The money rolled in.

McCormick's big opportunity came as Iraq descended into sectarian violence after 2003. His bomb detectors were avidly sought after by the Iraqi authorities, which bought 5,000 of the latest versions, the ADE 651s, and used them at road blocks to try to detect explosives in cars. They never worked, and the false reassurance they gave almost certainly contributed to many deaths. On the profits McCormick bought a £3m house in Bath, a house in Cyprus, a yacht and a string of horses. Eventually, after journalists investigated, he was sentenced to ten years in jail, still protesting that the devices worked because of 'nuclear quadropole resonance theory', whatever that is.

The tale of the fake bomb detectors is disturbing because there is no doubt that it was a fraud, yet a thin patina of 'innovation' managed to turn it into something just plausible enough to sell. People wanted to believe that it would be possible to detect bombs with a simple device so they fell for a fake innovation. McCormick cleverly put a high price on his device: a cheaper price might have given the game away.

Phantom games consoles

Only slightly less deceptive is the habit of announcing an innovation before it is ready, perhaps even when you know it is never going to be ready. An entrepreneur named Tim Roberts founded a company called Infinium Labs in 2002. It later changed its name to the apt Phantom Entertainment. The company promised to make a 'revolutionary new gaming platform' allowing on-demand video games online, rather than relying on cartridges or disks to load the games. It would be able to play both current and future personal computer games. Eager game users looked forward to the launch of the product in 2003.

In August 2003 the company announced a delayed date of early 2004 and a price of $399. The date was then postponed to November 2004. Then January 2005. Then March. Then September. In August 2006, the company simply dropped all mention of the product from its website altogether. By that time the Securities and Exchange Commission had accused Phantom and Roberts of illegally boosting the share price with, ahem, phantom announcements. Roberts paid a fine and agreed to a ban on being a company director as part of a settlement with the SEC.

This was an archetypal case of 'vapourware', namely software announcements that just kept evaporating and in some cases were probably designed and timed to entice customers into not buying competitor products. The word had been coined in 1983 by Esther Dyson. The slightly more benign version, namely that you announce before you have finally cracked a problem that you are confident you will crack, is much older. This is known as 'fake it till you make it' after an (unrelated) psychological tip often given to people to help them be more confident. Thomas Edison was not averse to announcing products that he could not yet make, including a reliable light bulb. And, to be frank, faking innovation till you can make it has stood some of the pioneers of the digital industry in good stead over the years. But it led to one of the big scandals of recent years.

The Theranos debacle

There was nothing flaky about Elizabeth Holmes as a teenager. She was an ambitious, hard-working, sleep-rationing school student who set about learning Chinese and getting experience in different biomedical laboratories even before she started at Stanford University. She came from a well-connected family but was determined from an early age to make her own way in her chosen field of medical diagnostics. In 2003 she dropped out of Stanford at nineteen to start a company, which became Theranos, laudably aimed at giving people pain-free, low-cost proactive healthcare as convenient as a smartphone, on the basis of a tiny drop of blood. She recruited her professor, Channing Robertson, and one of his PhD students to join her, while attracting an initial $6m in venture capital from some big names in Silicon Valley, all on the strength of her charisma. The business plan was to do blood tests simply and efficiently. It was a sure-footed start to a promising entrepreneurial career.

At the heart of her plan was a patented innovation: a patch containing microneedles for drawing blood and a silicon chip to do the analysis and create a disease map for each individual. The patch did not yet exist even in prototype, let alone work, nor did the chip, but at the rate things were changing in Silicon Valley it was plausible that they soon might. Holmes was essentially betting on Moore's Law delivering for her: she would fake it till she could make it. Her hero was Steve Jobs, who had conjured miracles from technology at Apple by demanding the apparently impossible and refusing to take no for an answer. She called her product 'the iPod of healthcare', wore black turtlenecks, sipped kale shakes and talked frequently of her admiration for Steve Jobs. 'Do or do not, there is no try,' she would say, citing Yoda from *Star Wars*.

But miniaturization did not prove so easy in microfluidics as in semi-conducting. Whereas a transistor became more reliable as it shrank in size, a blood-diagnostic test became less so. Holmes soon dropped the patch and went for a slightly more realistic idea of a cartridge, in which a small amount of blood, drawn from a fingertip, would be drawn into a patented 'nanotainer' then separated, tested against specific reagents and the results transmitted to a lab. She developed a lab-based robot called Edison, then a shrunken version called a miniLab, which contained a spectrophotometer, a cytometer and an isothermal amplifier. She aimed to take on and disrupt the lucrative duopoly of firms that dominate the blood-testing industry.

But none of the devices ever quite worked, a fact Theranos somehow kept under wraps, even from many employees. A stream of people left the firm, disillusioned or sacked. Lawsuits were filed against rivals and patent infringers, including against a family friend of Holmes's parents, who combined a career in the CIA with another in medical-device innovation, Richard Fuisz. This case eventually led to the suicide of Theranos's chief scientist, Ian Gibbons, who was responsible for many of the inventions patented

in Holmes's and his name, who had been demoted for voic-
ing his concerns about the firm's technology. He took an
overdose on the eve of being deposed in the Fuisz patent-
infringement suit.

Even though the innovations failed to deliver, let alone
match Holmes's ambitions for them, Theranos became
the darling of Silicon Valley. On the board sat a galaxy
of elderly political stars, eventually including Secretaries
George Shultz, William Perry and Henry Kissinger, Senators
Sam Nunn and Bill Frist, and General Jim Mattis. None
of these great and good names knew a thing about microfluidics, but their presence mightily impressed potential
customers. In 2011 Theranos did a deal with Walgreens to
put machines in Walgreens stores to carry out 192 instant
tests on customers' blood, using mainly chemiluminescent
immunoassays. Fear of missing out on an innovative technology drove Walgreens executives to override the concerns
of the very expert they had hired to check out Theranos's
claims. Likewise, the supermarket chain Safeway partnered
with Theranos to test the blood of staff in preparation for
launching wellness centres for customers. When Safeway
managers grew suspicious that test results were slow and
unreliable, their worries too were dismissed by senior Safeway executives who had been charmed by Holmes.

Around the same time, a military contract to use Theranos blood tests in the field, resulting from a persuasive
conversation that Holmes had with General Mattis,
led to questions being asked by experts at the Pentagon
about the regulatory status of Theranos's devices. Holmes
complained at this cheek to Mattis, who carpeted the regulatory officials. The project then stalled as Theranos failed
to deliver. None the less, Theranos boasted that its devices

were in use on the battlefields of the Middle East. It also claimed that Johns Hopkins Medical School had conducted due diligence on Theranos's technology and verified that it was 'novel and sound', whereas in fact the company had not even supplied a device to Johns Hopkins. Again and again, those who asked to visit Theranos's laboratories were fobbed off with excuses or shown labs with only conventional blood analysers made by other companies.

Yet still the money and the celebrity endorsements poured in. Those who invested more than $100m each included the Walton family, Rupert Murdoch and the future US education secretary Betsy DeVos. By 2014 Theranos was valued at an astonishing $9bn – more than Uber – while Elizabeth Holmes was now a billionaire, had graced the cover of several business magazines and had been profiled in the *New Yorker*. President Barack Obama appointed her an ambassador for global entrepreneurship; Bill Clinton interviewed her on stage at a Clinton Foundation conference; she hosted a fund-raiser for Hillary Clinton; Vice-President Joe Biden opened her new laboratory, saying: 'I know the FDA recently completed favorable reviews of your device', which was not quite true. Because Theranos was planning to use, but not sell, its devices, they fell through a crack in the federal regulations. The general view investors, directors, clients and commentators adopted was that somebody else must have checked out that her innovations worked, otherwise she could not possibly have been so successful in raising funds: a rather circular argument.

Quite what Holmes and her deputy (and secret romantic partner) Sunny Balwani thought would eventually happen is not clear. Perhaps they expected a real breakthrough in the technology to save them. But by hoping that the microfluidics breakthrough would happen, they were breaking a key rule of innovation, to tackle the most difficult issue first, in case it's insoluble. Google's 'X' team, which specializes in crazy 'moonshot' innovation schemes, calls this the 'monkey first': if your project aims to have a monkey recite Shakespeare while on a pedestal, it's a mistake to invent the pedestal first and leave till later the hard problem of training the monkey to speak.

Or perhaps Holmes and Balwani deceived themselves into thinking the breakthrough had already happened and was just prevented from emerging from the laboratory by the incompetence of the employees they kept firing. It pays not to underestimate self-deception and noble-cause corruption: the tendency to believe that a good cause justifies any means. As Nicole Alvino, a veteran of the Enron scandal, wrote, in relation to the Theranos story: 'A scam doesn't come about in one single moment. Rather, its creation is more like a slow trail of breadcrumbs, the end result of many little, seemingly innocuous decisions made along the way.' Like almost anything complex, crimes evolve.

Either way, the Theranos story is a spectacular case of failed innovation. The world has grown so used to miraculous and disruptive changes wrought by innovation that it sometimes forgets to be sceptical about wild claims backed by hubris.

Eventually, one of Theranos's laboratory directors resigned from the company and nervously blew the whistle on what was happening there, first to a blogger and then to an investigative reporter from the *Wall Street Journal*, John Carreyrou. He told Carreyrou that a culture of fear ruled the company, and that it was using Siemens machines for most of its tests, diluting the blood samples to create enough liquid to get a result, which in itself made the results less reliable. The tests it was doing on its own Edisons, such as for thyroid-stimulating hormone, were giving crazy results. Regulators were being duped and shown only one of the two laboratories. The company was breaking proficiency testing rules.

Worst of all, people were being told they were healthy when they were not, and vice versa. Carreyrou quickly confirmed this by having himself tested and getting four false alarms about his health, all soon contradicted by more conventional analyses. Theranos responded to Carreyrou's inquiries with threats and intimidation of his known and presumed sources by a phalanx of expensive lawyers, while refusing him an interview with Holmes.

When Carreyrou's story broke in the *Wall Street Journal* in October 2015, Holmes reacted with furious denials. His claims were 'actually and scientifically erroneous and grounded in baseless assertions by inexperienced and disgruntled former employees and industry incumbents', Theranos claimed. One of Carreyrou's sources, Tyler Shultz, refused to buckle despite coming under intense pressure from Theranos lawyers and his own grandfather, George Shultz, and despite spending $400,000 of his parents' money on lawyers' fees. More sources now came forward. Rupert Murdoch also refused to rein in the *Wall Street Journal*, despite strong pressure from Holmes and the Theranos board to do so, and despite his massive investment.

Finally, following the revelations, federal regulators acted against Theranos, finding serious deficiencies in laboratory practices that risked 'immediate jeopardy to patient health and safety'. In 2017 Theranos settled several lawsuits brought by investors. On 14 March 2018 the Securities and Exchange Commission indicted Theranos, Holmes and Balwani on civil charges of 'an elaborate, years-long fraud'. On 14 June Holmes and Balwani were indicted on nine counts of criminal wire fraud and two counts of conspiracy to commit wire fraud. They pleaded not guilty and a trial is set to begin in 2020.

By the time it was stopped, Theranos had tested almost a million people's blood, almost certainly giving both false alarm and false reassurance to a large number of people. It was about to roll out its service on a far greater scale through more than 8,000 Walgreens stores. John Carrey-rou's investigation therefore almost single-handedly prevented a health catastrophe. He argues that a general lesson still needs to be learned: 'hyping your product to get funding while concealing your true progress and hoping that reality will eventually catch up to the hype continues to be tolerated in the tech industry.'

Today other firms are claiming to have achieved at least some of what Theranos aimed to do. Sight Diagnostics, an Israeli company, is using machine vision to identify cells in blood to diagnose various disorders, including malaria, using a finger-prick of blood. But the Theranos debacle makes it hard for such companies to be taken seriously. An innovation fiasco can leave scorched earth behind.

Failure through diminishing returns to innovation: mobile phones

Most innovation failures are not fraudulent. Many come from honest attempts to improve the world that don't quite achieve their aims. As an example, consider the history of the mobile-phone market.

From the moment when mobile telephones became small and cheap enough to catch on in the 1990s they experienced continuous innovation. The handsets shrank, the batteries slimmed, reliability improved and new features exploded. Text arrived with Nokia in 2000. Motorola incorporated the camera in 2005. Blackberry gave us mobile emails in 2006. The iPhone brought the touch screen, music and apps in 2007. The smartphone partly or wholly replaced the need to own a camera, a flashlight, a compass, a calculator, a notebook, maps, address books, filing cabinets, television, even a pack of cards. By 2016 we were watching movies, sharing selfies and surfing social media on Samsung Galaxies and iPhone 6s. No longer black and functional, mobiles became colourful and sleek. After shrinking steadily from the bricks of the early 1990s, smartphones began growing larger again, though flatter. Change was incessant. Upgrading to the next machine seemed as natural as changing clothes fashions.

Yet Nokia, Motorola and Blackberry all fell painfully to earth. Nokia had started out in 1865 as a pioneer of papermaking from wood at a water mill, then become an electricity generator, then turned to products for forestry workers such as boots, before plunging early and brilliantly into mobile phones. The firm spent $40bn on research and development over a decade from 1992, far more than Apple, Google or anybody else in the industry. It could afford to, because Nokia was worth more than $300bn in 2000 and by 2007 it had 40 per cent of the entire global market in mobile phones. The money spent on research and development came up with the right ideas: early prototypes of smartphones and tablets, with colour touch-screens above a single button, just like Apple. But the company failed to develop the ideas into practical products, because of corporate caution, internal fights between rival teams of software engineers and the dominance of the voice-phone sector within the firm. Nokia thought it had time to shift to a world in which mobile was all about software not hardware: it wanted to move smoothly, not suddenly, out of its core business. Qualcomm's chief executive, Paul Jacobs, found that Nokia spent much more time than other device makers thinking before doing: 'We would present Nokia with a new technology that to us would seem as a big opportunity. Instead of just diving into this opportunity, Nokia would spend a long time, maybe six to nine months, just assessing the opportunity. And by that time the opportunity often just went away.' Nokia, like Blackberry, failed to see just how revolutionary and popular the iPhone would be, despite what they saw as its obvious limitations. Microsoft eventually bought Nokia's handset business for just $7.2bn. Innovation, more often than not, eats its own offspring.

Around 2017–19 innovation in smartphones began to stall. People saw less need to upgrade and sales began to falter. Annual global sales had been heading inexorably for two billion units a year, but never got there and probably never will. The new features offered seemed only marginally useful and the price that came with them exorbitant. Moving up to 3G was a must; going to 4G felt like a maybe; future 5G feels like a luxury, and anyway it is coming more slowly than expected. In 2019 Huawei launched the Mate X, which could fold out into an eight-inch square but cost a crazy $2,600. Samsung also unveiled the Galaxy Fold, but the screen kept breaking on the prototypes so the commercial launch was delayed. This is innovation failure not because of fraud or fakery but because of diminishing returns. There is a limit to what seems useful from a pocket device. It turns out that – contrary to what a lot of people think – innovation cannot push new ideas on people unless they want them.

A future failure: Hyperloop

In 2013 Elon Musk, the founder of the Tesla car company, published a manifesto for a new transport system he called Hyperloop. He said existing plans for high-speed rail between cities were relying on old, costly and inefficient technologies, and that it was time to seek out a new mode of transport that would be safer, faster, cheaper, more convenient, immune to bad weather, 'sustainably self-powering', resistant to earthquakes and not disruptive to those along the route. The answer, he suggested, was a tube, containing a partial vacuum, through which twenty-eight-person buses (or 'pods') would zoom at up to 760 mph on magnetic levitation, driven by solar power backed up with lithium batteries. It would take thirty-five minutes to travel from Los Angeles to San Francisco and the whole thing would cost $7.5bn to build, or about one-tenth of high-speed rail, promised Musk.

He is right that air resistance is the biggest limiting factor on rapid transport, while friction on wheels is another big drain, which is why planes go high into the thin air of the stratosphere and missiles hop into space. And planes are weather-dependent to some degree. But creating thin air at ground level is costly and difficult, as is magnetic levitation, while accelerating and going round curves at high speeds inside a tube is risky and requires careful engineering, not to mention mostly straight lines or very gentle curves.

Musk did not propose to build the hyperloop himself but threw it out there as an open-source idea for others to progress. Within a few months various startups were running with the idea in America, China, Europe and elsewhere. Hyperloop Transportation Technologies has built a five-mile test track in Quay Valley, California. Most of the entrepreneurs pursuing this dream now think of almost pure vacuums, rather than the air fans that Musk suggested. Consultants have polished PowerPoint presentations for gullible investors all over the world. For the truth is that the idea of hyperloops being cheaper or more reliable than existing transport methods is for the birds.

First, it is not a new idea at all. As early as 1800 George Medhurst patented an 'Aeolian engine' to propel coaches using air pumps, and by 1812 he had a 'plan for the rapid conveyance of goods and passengers upon an iron road through a tube of 30 feet in area by the power and velocity of air'. In 1859 the London Pneumatic Despatch Company was formed to build a railway that would send packages between post offices through an underground tube, at speeds of up to 60 mph, using compressed air driven by a steam engine. By 1865 it had opened its first commercial line, between Euston and Holborn, and to celebrate it sent the company's chairman, the Duke of Buckingham, through the tube in one of the special capsules. Financial and engineering problems ensued. Parcels became stuck in the tubes (luckily the duke avoided this fate). By 1874 the Post Office had abandoned the system and the company soon went into liquidation.

Pneumatic passenger lines were tried from time to time but fared no better. Alfred Ely Beach designed and built a one-car, one-stop atmospheric train to run under a street in

Manhattan in 1870. It was intended as a demonstration, with passengers being carried to one end of the line and back to alight where they started, inside cars pushed by the force of air. By the time it closed 400,000 people had ridden it, but the idea never took off.

These nineteenth-century schemes did not include the idea of a vacuum, but even that idea is fairly old. In 1910 Robert Goddard came up with exactly the same plan as Musk: a magnetically driven train in an evacuated tube to travel from Boston to New York in twelve minutes. It remained on the drawing board. In the 1990s and the early 2000s, various people proposed magnetic-levitation trains in vacuum tubes. So there is nothing new in the concept, and there is no reason to think a breakthrough in some technology has transformed things. Transport technologies have not experienced Moore's Law, probably because the things to be transported, people, have not become smaller.

Next consider the engineering issues. If you build a tube strong enough to contain a vacuum that is hundreds of miles long, don't expect it to be lightweight. It would need strong walls and substantial foundations to keep it straight and level. It would need flexible thermal expansion joints to cope with warm days after cold nights, and these must be airtight, yet strong enough to withstand the 14.7 pounds per square inch of atmospheric pressure on every square inch of its vast surface.

Maintaining the vacuum would not be easy, and in an emergency there would have to be a mechanism to return the tube to atmospheric pressure to rescue the passengers. But any such mechanism would risk leaks. Re-pressuring and re-evacuating the tube would take time. The pods themselves would have to be pressurized, and would load passengers at atmospheric pressure before entering the vacuum through an air lock, which might malfunction. None of these problems are insuperable, but – remember the golden rule of innovation – overcoming them will be bound to require trial and error, not just clever forecasting, and may not be cheap.

Then there is the land issue. The hyperloop will either have to be in a tunnel, which is expensive to drill, or on struts above the ground, but land does not come cheap and nor is it easy to find a straight route above ground that does not go through people's houses, over roads or rivers and through hills. (Ask railway and road builders.) Quite why it should be cheaper to build this than a railway is never clear.

Then there is the energy requirement. A modern magnetic-levitation train such as Japan's Chuo Shinkansen uses more, not less, energy than one on rails. Putting one in a vacuum saves energy, but brings a penalty too, because not only do the vacuum pumps need power, but since air resistance cannot help to slow the train, braking requires more energy. Musk envisages all the energy coming from solar power, but this is still expensive – despite the falling costs of solar panels themselves – because of the cost of land, infrastructure and maintenance. A lot of land would be needed for the solar farms and a lot of batteries to support night-time schedules.

On top of this, the capacity of the line would be limited. To carry 5,000 passengers an hour, a hyperloop would need to send 180 twenty-eight-man pods an hour through the tube. That is three departures every minute. At the very least you would have to arrive early and queue for a long time for your hyper-punctual pod departure. Logistically, this would make a busy airport look simple, especially if different pods were to branch off to different destinations. I have not yet mentioned security, but it would be just as tight as at airports.

Although determined innovators may solve some of these problems, there is no guarantee they will crack them in a way that saves money and makes hyperloops competitive with rail or flight. If cars and trains and planes did not exist, then even an inefficient hyperloop transport would be worthwhile. But they do exist and it would have to earn its living in competition with these long-established modes of transport. It is hard not to conclude the hype over hyperloop is because we have come to believe innovation can solve almost any problem.

Failure as a necessary ingredient of success: Amazon and Google

If the world rules out all innovation failure as fraud, or takes too cautious an approach, then it will stop innovation in its tracks, as many a country and company have experienced. The central theme of innovation, after all, is trial and error. And error is failure.

Take the case of the millennium bridge in London, a footbridge built across the Thames and opened with much fanfare. Designed to have a shallow profile, like a blade, it was hailed as a fine addition to London's riverscape. Over 90,000 people crossed it on the first day, 10 June 2000, and almost immediately a problem became apparent. As more people walked its length, the bridge began to sway, very slightly at first, from side to side. The swaying increased. The bridge was closed on its opening day and then reopened for restricted numbers of people, but the problem recurred.

After two days, the now infamous 'wobbly bridge' was closed for a year and a half while work costing £5m was done to stabilize it. It turned out that a very slight tendency to sway from side to side was being reinforced by people's instinctive reaction to the swaying, in a case of positive feedback: the more it swayed, the more people moved in such a way as to make it sway. After thirty-seven dampers were installed, it reopened successfully and is now a routine part of London's infrastructure.

Amazon is a good example of failure on the road to success, as Jeff Bezos has often proudly insisted. 'Our success at Amazon is a function of how many experiments we do per year, per month, per week. Being wrong might hurt you a bit, but being slow will kill you,' Bezos once said: 'If you can increase the number of experiments you try from a hundred to a thousand, you dramatically increase the number of innovations you produce.'

Having cleverly spotted that books were good candidates for online retail, rebuffed the challenge of the big bookstores' online efforts to snuff him out and launched the company's public share offering in 1997, Bezos then set out to become the head of a general technology company with fingers in every internet pie, and to get big fast. Amazon raised over $2bn in the dotcom boom between 1998 and 2000 and spent most of it on acquiring dotcom startups. It bought a market site called Exchange.com, a social network site called PlanetAll, a data-collecting company called Alexa Internet, a film database called IMDB.com, a British bookseller called Book Pages and a German online book store called Telebuch. As Brad Stone documented in his book *The Everything Store*, Amazon also sprayed venture-capital money at other startups: Drugstore.com,

Pets.com, Gear.com, Wineshopper.com, Greenlight.com, Home-grocer.com and Kozmo.com. Almost all of these failed in the bust that followed.

In 1999 Amazon wrote off $39m in unsold merchandise after an ill-fated venture into toy retailing. It launched Amazon Auctions, which failed to compete with eBay. The chief operating officer, Joe Galli, resigned, the stock price fell and anxiety pervaded the company. Stone describes the mood: 'As the new millennium dawned, Amazon stood on the precipice. It was on track to lose more than a billion dollars in 2000.' A stock analyst at Lehman Brothers, Ravi Suria, accused Amazon of an 'exceedingly high degree of ineptitude' and predicted the firm would run out of cash within a year. The share price continued to fall. In 2001 the company laid off 15 per cent of the workforce. Had Amazon failed at that point or a little later – and even as late as 2005 eBay was worth three times its value – it would have been a cautionary tale of hubris and nemesis.

But what Suria saw as ineptitude was actually an appetite for experiment and a tolerance for failure. Among the initiatives that went wrong were always some that went right. Again and again Bezos found himself fighting his Amazon colleagues for an idea they thought was rubbish. One was to cease spending money on advertising. Another was the launch of Marketplace, whereby third-party sellers of produce could compete with Amazon itself. 'As usual it was Jeff against the world', as one colleague put it. Bezos's management style was specially designed, he hoped, to avoid the institutionalized middle-management complacency that soon stifled innovation at large firms, including Microsoft. Hence his tendency to hire people into small 'two-pizza' entrepreneurial teams often in competition with each other, his allergy to big meetings and PowerPoint presentations and his operation of a sort of reverse-veto policy, whereby a new idea has to be referred upwards by managers even if all but one of them thinks it is rubbish. All of these were designed to encourage innovation, and effectively to allow failure to happen but relatively painlessly. It was this sort of Darwinian process that led Amazon to the discovery of an even bigger business than online retail, namely the provision of cloud computing to outsiders, which became Amazon Web Services. Google and Microsoft were slow to spot what Amazon was doing, and how much it enabled tech startups to get going. Bezos, like Edison in the nineteenth century, understood that transformative, disruptive innovation is not a matter of making a new toy, but of launching a new business built around the needs and wants of real customers. And to find that holy grail requires a lot of honest failure along the way.

Google, likewise, tolerates and even encourages failure. Its moonshot subsidiary, known as 'X', launched in 2009, sets out to find big, disruptive new business opportunities. Most of these fail. The high-profile launch of Google Glass, a miniature screen and voice-activated camera attached to a pair of eyeglasses, was X's most public and most expensive misjudgement. Google launched the product in April 2013 for pioneering 'Glass Explorers' to use, and to the public a year later at a cost of $1,500. Just seven months later the company stopped selling the product, promising to bring it back within two years. It never did. What went wrong? Customers baulked at the price, the risks – to health and privacy – and the lack of any useful purpose being served. This was innovation for the sake of it, they concluded, and it added nothing to their lives, or at least not $1,500 worth.

Google still pursues specialized uses for the technology in hospitals and other settings, but as a consumer product it was a failure. If Google Glass had been a government project, the chances are they would still be ploughing on with it.

Project Loon, which wanted to put Wi-Fi on balloons, came unstuck when it emerged that the balloons could not be prevented from leaking. A project called Foghorn, to extract carbon dioxide from seawater, combine it with hydrogen, also extracted from water using electricity, and make fuel from reacting the two, was another of X's projects. It sounds a bit like a liquid perpetual motion machine: after all, the laws of thermodynamics suggest that turning combusted products (CO_2 and H_2O) back into combustible ones will need more energy than it can deliver. But so keen was X on grappling with the seemingly impossible that it was even prepared to have a crack at the laws of thermodynamics. Kathy Hannun of X pulled the plug on Foghorn in 2016 when she realized they were never going to get to the goal of $5-a-gallon fuel, let alone within five years. Such ruthlessness is crucial to the incubation of experiments. But Astro Teller, the head of X, celebrates rather than laments such failures. In 2016 at TED in Vancouver he spoke of the 'unexpected benefit of celebrating failure'. One day, X will perhaps generate something so spectacular that it dwarfs Google itself.

Lockheed Martin pioneered this idea of a high-risk company within a company, licensed to try crazy things in case some of them led to immense rewards. It opened its secret Advanced Development Programs, better known as the 'skunk works' in 1943 and produced some of the first jet fighters and high-altitude spy planes. Bell Labs, a subsidiary of AT&T, operated in a similar fashion from the 1920s, and invented all sorts of new technologies, including the transistor and the laser, but gradually became more of a science lab than a tech one, winning eight Nobel Prizes. Xerox's Palo Alto Research Center also proved a valuable laboratory for applying new ideas and incubating new businesses.

A high appetite for failure is crucial to these skunk works, and there seems to be something about the West Coast culture that allows this to happen more easily. Kodak, Blackberry, Nokia and many other firms, based elsewhere in the world, failed to replicate this appetite for useful failure. There is a case to be made that what makes the West Coast special in this respect is the legality of its dual-share ownership structure, whereby the founders retain voting control of the company while investors merely get to enjoy the ride. That way the founders can take risks, make long-term bets and ignore at least some of the impatience or caution of their shareholders. But it is clearly a self-reinforcing thing, and has taken root, psychologically, in Silicon Valley more than anywhere.

In his book *Non-Bullshit Innovation*, David Rowan recounts the extraordinary story of Naspers, a conservative South African newspaper-publishing company that rigidly supported the cause of Afrikaner nationalism for decades, before pivoting successfully into technology investing in the 1980s. At the suggestion of a brash young man called Koos Bekker it built Africa's first cable-television network in the 1980s, then its first mobile-phone network in the 1990s. Neither venture was easy or cheap, and both were high-risk gambles, but both paid off in the end. Then, like Nokia, Naspers stumbled, losing $400m on one venture in Brazil, followed by a string of expensive internet duds in China, one of which cost Naspers $46m in just six months.

There the story might have ended, showing that luck runs out for gamblers in the end. But it did not. With one last throw of the dice, Bekker changed tack, seeking out a promising Chinese-owned startup, rather than trying to start one of his own in China. He stumbled on a little enterprise called Tencent, run by Pony Ma, son of the Shenzhen harbour master, which had somehow signed up two million instant-messaging customers in the city but was not sure how to get any revenue out of them. In 2001 Bekker put $32m into Tencent for a 46.5 per cent stake. Seventeen years later that stake was worth $164bn.

11

Resistance to innovation

When a new invention is first propounded in the
beginning every man objects and the poor inventor
runs the gauntloop of all petulant wits.

WILLIAM PETTY, 1662

When novelty is subversive: the case of coffee

Innovation is the source of prosperity and yet it is often
unpopular. Take the case of coffee, a late arrival in civilization, not reaching Europe or Asia much before the 1500s.
Coffee is an Ethiopian plant whose beans can be roasted to
make the basis of a stimulating, addictive drink. Because
grinding and roasting the beans just right requires machinery, coffee tends to be bought and drunk in public spaces.
The spread of chains of coffee shops around the world with
their fancy menus and reputations as ideal meeting places
is a widespread modern phenomenon, associated with Star-
bucks. Yet that is just the latest version of the pattern. Coffee
shops have been popular places to meet for four centuries.
In 1655 an apothecary named Arthur Tillyard founded

the Oxford Coffee Club for students to discuss ideas over hot drinks at what might be called his 'café'. Seven years later the club became the Royal Society, Britain's science academy.

Yet the history of coffee also shows a key feature of innovation: that it almost always meets resistance. As coffee spread to Arabia, Turkey and Europe in the 1500s and 1600s, it encountered fierce opposition and was frequently – if in the end ineffectually – banned. In 1511 Kha'ir Beg closed down the coffee houses of Mecca, burning all supplies of the beans and beating those caught with them. He was over-ruled by the Sultan of Cairo, but not for long. In 1525 Mecca's coffee houses were banned again. By 1534 the opposition to coffee had reached Cairo, where the mob attacked coffee houses. Here the ban failed and coffee persisted. It even became a law that if a man failed to provide coffee for his wife, that was grounds (groan) for divorce.

Coffee reached Constantinople in the 1550s and was promptly banned by Sultan Selim II. It was banned again in 1580 by the usurper Murad III and again in the 1630s by Murad IV. This implies that the bans failed each time, but why were these rulers so keen to stamp out this drink? Mainly because coffee houses were places of gossip and therefore potential sedition. Murad III was paranoid that the fact that he had killed his whole family to claim the throne might have been a topic of conversation in coffee houses. I dare say he was right and the subject did sometimes crop up.

In 1673 King Charles II of Scotland and England tried to ban coffee houses and was splendidly honest about why he was such a keen prohibitionist:

As for coffee, tea and chocolate, I know no good they do; only the places where they are sold are convenient for persons to meet in, sit half day and discourse with all companies that come in of State matters, talking of news and broaching of lies, arraigning the judgments and discretion of their governors, censuring all their actions, and insinuating into the ears of the people a prejudice against them; extolling and magnifying their own parts, knowledge and wisdom, and decrying that of their rulers; which if suffered too long, may prove pernicious and destructive.

Yet there were other reasons to resist coffee too. People who made and sold wine in France, or beer in Germany, opposed this new competitor, one that stimulated rather than anaesthetized the customer. In Marseilles in the 1670s, the vintners found allies in the medical profession, especially at the university of Aix, where two professors commissioned an attack on coffee by a medical student known only as 'Colomb'. His pamphlet, entitled 'Whether the Use of Coffee is Harmful to the Inhabitants of Marseilles', argued that the 'violent energy' of coffee entered the blood, attracted the lymph and dried the kidneys, rendering people exhausted and impotent: pseudoscientific nonsense, of course, bought and paid for. Around the same time in London there was a battle of pamphlets on the topic. 'A Broad-side against Coffee, or the Marriage of the Turk' in 1672 was answered two years later by one from the founder of London's first coffee house, a Lebanese trader by the name of Pasqua Rosee: 'A Brief Description of the Excellent Vertues of that Sober and Wholesome Drink called Coffee'.

As late as the second half of the eighteenth century, Sweden tried to ban coffee no fewer than five times. The regime confiscated coffee cups from its citizens in a desperate effort to enforce the ban, and ceremonially crushed a coffee pot in 1794. King Gustav III set out to prove coffee was bad for people through a controlled experiment. He ordered one convicted murderer to drink nothing but coffee while another drank nothing but tea. Magnificently, both men outlived the doctors monitoring the experiment, and even the king himself. The coffee-drinking murderer lived longest of all, of course. Campaigns against coffee none the less continued in Sweden until the twentieth century.

Here we see all the characteristic features of opposition to innovation: an appeal to safety; a degree of self-interest among vested interests; and a paranoia among the powerful. Recent debates about genetically modified food, or social media, echo these old coffee wars.

In his book *Innovation and Its Enemies*, Calestous Juma tells the story of the coffee wars, and also that of the margarine wars. Margarine, invented in France in 1869 in response to the rising price of butter, was subjected to a decades-long smear campaign (Professor Juma's pun, not mine) from the American dairy industry, not unlike recent campaigns against biotech crops. 'There never was, nor can there ever be, a more deliberate and outrageous swindle than this bogus butter business,' thundered the New York Dairy Commission. Mark Twain denounced margarine. The governor of Minnesota called it an abomination. New York state banned it. In 1886 Congress passed the Oleo-margarine Act to restrict its sales through burdensome regulation. By the early 1940s, two-thirds of American states still had bans on yellow margarine on their books, on spurious health grounds. The National Dairy Council campaigned against margarine, inventing evidence as it did so: a university experiment it reported in which two rats were treated like the Swedish murderers, one fed margarine and one butter, with terrible results for the health of margarine rat, turned out to be entirely made up. But the margarine industry was not a passive victim. Indeed, the origin of the long-running theory that dietary fat was the cause of heart disease, now largely debunked, lies in studies in the 1950s stimulated partly by the vegetable-oil industry in its fight-back against the butter industry.

Juma chronicles how hansom cab operators in London furiously denounced the introduction of the umbrella; how obstetricians long rejected the use of anaesthesia during childbirth; how musicians' unions temporarily prevented the playing of recorded music on the radio; how the Horse Association of America for many years fought a rearguard action against the tractor; how the natural-ice harvesting industry frightened people with scares about the safety of refrigerators. Truly, there is likely to be a backlash against any new technology, usually driven partly by vested interests but clothing itself in the precautionary principle. In 1897 one London commentator worried that the telephone would destroy private life if it was not restricted: 'we will soon be nothing but transparent heaps of jelly to each other.'

When innovation is demonized and delayed: the case of biotechnology

The campaign to prevent the spread of biotechnology in European agriculture echoes the campaigns against coffee and margarine, but with more lasting success – so far. The two key weapons are demonization and delay: making claims of danger and demanding delay to implementation in the hope that this deters business investment.

The development of transgenic crops, mainly in the United States in the 1990s, went smoothly at first. Some opposition was stirred up, but not much. A dress rehearsal a decade earlier over genetically modified bacteria used to prevent freezing of strawberries had petered out. But when GMOs came to Europe, suddenly everything changed.

It was indeed sudden. In 1996 a campaign in Britain to force supermarkets to label genetically modified foods failed for lack of interest. Yet by 1999 biotechnology was in full retreat before an army of activists and critics, with big money behind them and prominent supporters, including politicians. The industry had by then all but abandoned attempts to grow trial crops, as each one attracted vandals in white boiler suits, some of them members of the House of Lords. A few years later the biotech industry effectively abandoned Europe altogether, and any attempt to genetically modify its main crops such as wheat.

What caused this sudden change? In March 1996 the British government acknowledged for the first time that there was a potential risk to human health from eating beef contaminated by bovine spongiform encephalopathy (BSE), better known as 'mad cow disease'. That very same month the European Commission approved the first import of genetically modified soybeans into Europe. The coincidence led to confusion between the two issues and a collapse in trust in all government reassurance about safety. In fact, few people were to die from BSE, and none from GMOs, but the damage was done. As Robert Paarlberg puts it: 'Efforts by European officials to reassure consumers about the soybeans had no impact, since the BSE case had destroyed their credibility as guardians of food safety.' Combining food safety concerns with an antipathy to big business, Greenpeace and Friends of the Earth, two very big campaigning businesses active in the United Kingdom and always in search of new issues, spotted from their market research that there was public disquiet about these new types of crop and that it could be lucrative to fan the flames.

Since 2005 Canada has approved seventy different transgenic varieties of crops while the European Union has approved just one, and that took thirteen years, by which time the crop was outdated. Mark Lynas, a prominent campaigner on the subject at the time who later changed his mind and became a strong advocate of genetic modification, remembers the heady days when juries refused to convict GM crop vandals arrested by the police and judges declared their admiration for the vandals; when GM ingredients were removed from school meals and supermarkets pulled them from their shelves; when the right-of-centre *Daily Mail* inveighed against 'Frankenfood'. It was not just in Britain. In France the activist-farmer José Bové destroyed rice crops and became a hero, while arsonists in Italy burned down a seed depot.

In response to this pressure, the EU placed a moratorium on all new GM crops. This later evolved into a system of regulatory approval that was so complicated and time-consuming that it amounted to a de facto ban. The EU had by now installed the precautionary principle as a guiding light. This superficially sensible idea – that we should worry about unintended consequences of innovation – morphed into a device by which activists prevent life-saving new technologies displacing more dangerous ones. As formally adopted by the European Union in the Lisbon Treaty, the principle holds the new to a higher standard than the old and is essentially a barrier to all innovations, however safe, on behalf of all existing practices, however dangerous. This is because it considers the potential hazards, but not the likely benefits of an innovation, shifting the burden of proof to an innovator to prove that its product will not cause harm, but not allowing that innovator to demonstrate that it might cause good, or might displace a technology that already causes harm. Thus organic farmers are free to use pesticides so long as they were invented in the first half of the twentieth century, even though the ones they do use, such as copper-based compounds, are significantly more harmful than modern ones – and not even 'organic' in any reasonable definition. Copper sulphate, for example, is described by the European Chemical Agency as 'very toxic to aquatic life with long lasting effects, may cause cancer, may damage fertility or the unborn child, is harmful if swallowed, causes serious eye damage, may cause damage to organs through prolonged or repeated exposure'. It also bio-accumulates, that is to say becomes more concentrated as it moves through the food chain from herbivores to carnivores. Yet it is repeatedly re-authorized by the EU, without any fuss of opposition, for use as a fungicide by organic farmers on human food crops, including potatoes, grapes, tomatoes and apples. The reason given is that no alternative pesticide is available to organic farmers. But this is simply because they choose to refuse safer, newer pesticides. This is a repeated pattern: the precautionary principle largely ignores the risks of existing technologies, defying the concept of harm reduction.

For Greenpeace and Friends of the Earth, this quick win against GMOs in Europe was a bit of a problem. The issue was a huge money spinner, so they needed to keep it going. They turned their attention to other parts of the world, gaining direct access to the negotiations for an international protocol on transboundary movement of living genetically modified organisms and drawing implausible parallels with the transport of hazardous waste. FoE then began attacking the United States for sending food aid to starving people in southern Africa, where a serious drought was in progress, and in 2002 succeeded in getting Zambia to reject GM maize destined for starving people. The pressure groups moved on to the rest of Africa and parts of Asia. Greenpeace turned its attention to blocking the humanitarian application of GMOs, especially in the form of beta-carotene-containing Golden Rice, specially designed by a non-profit project to prevent malnutrition and death among poor children. Golden Rice was developed by the Swiss-based scientist Ingo Potrykus and colleagues, in a long and laborious endeavour during the 1990s, purely as a humanitarian, non-profit project designed to alleviate the high mortality and morbidity caused by vitamin-A deficiency in people who rely on rice for food. By one estimate vitamin-A deficiency kills 2,000 mainly urban children under five every day, 700,000 a year, in countries where poor people eat rice as a staple diet: it lowers their immune resistance and causes them to go blind. Yet using every means at its disposal Greenpeace chose to campaign hard to block a technology that could prevent these deaths. At first Greenpeace argued that Golden Rice was useless at curing vitamin-A deficiency because the first prototype of the plant, which contained a daffodil gene, had too little beta carotene to be any use. It then switched to arguing that subsequent varieties of the rice, with a maize gene, had too much beta carotene and could be poisonous. Desperate to kill a potential good news story coming out of biotechnology, Greenpeace continued to lobby hard against the crop even as it was proved by relentless experiments to be safe and effective. It was in response to this shocking campaign that 134 Nobel Prize-winners called on Greenpeace in 2017 to 'cease and desist in its campaign against Golden Rice specifically, and crops and foods improved through biotechnology in general' (150 have now signed the letter). Their call was in vain.

Demonization of biotechnology led to a vicious circle as far as the companies involved were concerned. The more the activists demanded regulation and caution, the more expensive it became to develop new crops, and the more therefore it became impossible to do so except within large companies. There was thus a strange symbiosis between big industry and its critics. At one point activists demanded that Monsanto make crops that could not survive beyond the first generation, so they could not run riot in the wild as 'superweeds' – an entirely false fear anyway since most crops are ill suited to being weeds. In response, Monsanto therefore explored the possibility of developing genetic variants that were incapable of breeding true. Although it did not develop these, activists immediately accused it of introducing 'terminator technology' to trap farmers into buying new seed each year. The charge stuck.

The biotech industry kept trying to get Europe to change its mind. After all, it was enthusiastically importing genetically modified soybean from the Americas as cattle food, so why not grow biotech varieties? In 2005 the European Food

Safety Authority approved a genetically modified variety of potato produced by the German company BASF. But the EU did not give it market approval, citing the precautionary principle. BASF complained to the European Court of Justice in 2008. The European Commission responded by commissioning another evaluation from EFSA in 2009. EFSA once again said the product was safe and the EU had to approve its use in 2010, five years after the initial application. Yet the Hungarian government then found a bizarre way to stop the product. It argued that the EU had based its approval on the first EFSA approval, when it should have cited the second one, even though it came to the identical conclusion. In 2013 the General Court of the EU upheld Hungary's complaint and annulled the approval. By then BASF had lost interest in banging its head against this precautionary brick wall, withdrawn its application, packed up its entire research into GM crops and moved it to America.

One way in which the precautionary principle works to prevent innovation is by making experimentation difficult in the period between prototype and practical application. In the case of Golden Rice, the principle required the developers to get special approval, with a vast and laborious weight of evidence, for each variety that was to be tested in the field. That meant that, just as is the case with nuclear power plant design, it was impossible to try lots of varieties to find the one best suited to farmers' needs, a normal practice in plant breeding. Sure enough, the one variety that was picked initially proved disappointing and the breeders had to go back and try another one, wasting several precious years in which yet more children died. If Thomas Edison had needed to get special regulatory approval for every one of the 6,000 plant samples he tested as a filament in a light bulb, he would never have found bamboo.

Mark Lynas's verdict on the GMO episode is unsparing: 'We permanently stirred public hostility to GMO foods throughout pretty much the entire world, and – incredibly – held up the previously unstoppable march of a whole technology. There was only one problem with our stunningly successful worldwide campaign. It wasn't true.' As with the opposition to coffee, it is now clear that the opposition to genetically modified crops was wrong both factually and morally. The technology was safe, environmentally beneficial and potentially good for small farmers. The anti-GM movement caught on amongst wealthy people with plentiful, cheap food. It was not pressing and relevant to their lives to increase crop yields. Those who paid the opportunity cost of the prohibition were the sick and starving who had no voice. Even the big pressure groups have tiptoed away from it in recent years. But the damage was done.

When scares ignore science: the case of weedkiller

The herbicide glyphosate, also known as Roundup, has come to be a cheap and ubiquitous method of weed control since its invention by a scientist at Monsanto, John Franz, in 1970. It has huge advantages over other weedkillers. Because it inhibits an enzyme found only in plants it is virtually harmless at normal doses to animals, including people, and because it decomposes rapidly it does not persist in the environment. It is far safer than the stuff it replaced, paraquat, which was sometimes used by suicides. Glyphosate has transformed agriculture by allowing farmers to control weeds chemically, rather than by the more ecologically harmful activity of ploughing: it has led to the no-till revolution. This is especially true where crops genetically modified to be glyphosate-resistant are grown.

In 2015, however, a World Health Organization body called the International Agency for Research on Cancer (IARC) came to the conclusion that glyphosate might be capable of causing cancer at very high doses. It admitted that by the same criteria, sausages and sawdust should also be classified as carcinogens, while coffee was even more dangerous (and unlike glyphosate is regularly drunk). The effect would be minuscule: Ben & Jerry's ice cream was found to contain glyphosate at a concentration of up to 1.23 parts per billion, so a child would have to eat 3 tonnes of it a day before any risk would be encountered. Food safety authorities in Europe, America, Australia and elsewhere had all studied glyphosate in depth and concluded that it was not a risk at normal doses. The German Federal Institute for Risk Assessment had looked at more than 3,000 studies and found no evidence of any kind of harm to animals.

It soon turned out that IARC's conclusion was based on a biased review of the evidence. As Reuters reported: 'In each case, a negative conclusion about glyphosate leading to tumors was either deleted or replaced with a neutral or positive one.' It emerged that the scientist who advised IARC on the matter also received $160,000 from law firms suing Monsanto on behalf of cancer victims. As David Zaruk of the Université Saint-Louis, Brussels, put it, the tactic of activists in these cases is to 'manipulate public perception, create fear or outrage by co-operating with activists, gurus and NGOs, find a corporate scapegoat and litigate the hell out of them'. If Europe were to ban glyphosate it would open up a litigation bonanza in America, where bounty-hunting law firms are constantly in search of the next tobacco-sized windfall.

This sort of activism, whether you approve of it or not, is a significant deterrent to innovation. Taken together with a similar backlash against neo-nicotinoid insecticide seed dressings, despite their obvious improvement over previous insecticides in safety and collateral damage to non-target species, this opposition to herbicides has contributed to a significant slowdown in research and development on crop protection products. If you think new chemicals are usually better than old ones, and that growing enough food to feed the world population on as little land as possible is a good idea, then that is a bad thing.

When government prevents innovation: the case of mobile telephony

In Chapter 5 I argued that most technologies come along at about the right time and could not have been introduced much earlier. A possible exception may be the mobile telephone. The history of cellular telephony is an extraordinary story of bureaucratic delay imposed by government at the behest of various lobbies, as the economist Tom Hazlett uncovered in a 2017 book *The Political Spectrum*. We could have had mobile phones decades earlier than we did.

In July 1945 J. K. Jett, the head of the US Federal Communications Commission, gave an interview to the *Saturday Evening Post* in which he said that millions of citizens would soon be using 'hand-held talkies'. The FCC would have to issue licences, but that 'wouldn't be difficult'. The reason for this optimism, he stated, was that a 'cellular' concept would revolutionize the technology: the transmitter's talkie would not have to connect all the way to the receiver's talkie, but just as far as the nearest radio mast, which would be connected by wire to the mast nearest the receiver. Users could seamlessly switch to new cells as they moved. This would save energy and restrict use of the spectrum to a local area, allowing greater bandwidth. Instead of a few hundred conversations happening at once on radio, hundreds of thousands of conversations could occur.

However, in 1947 the very same FCC rejected AT&T's application to start a cellular service, arguing that it would be a luxury service for the few. Television was the priority and got the lion's share of the spectrum. 'Land mobile', the category that included cellular, got just 4.7 per cent. Yet television never used more than a fraction of its allocated frequencies, leaving what Hazlett calls a 'vast wasteland' of unused spectrum, 'blocking mobile wireless for more than a generation'. More than two-thirds of channels went unused in the 1950s, but broadcasters lobbied to defend their right to this empty territory, if only to stymie competition for the existing oligopoly of licensed television networks.

Mobile-telephone operators, called 'radio common carriers' (RCCs), did exist to service big firms like airlines or oil companies with offshore rigs, but not in cellular form, and were strictly limited to two operators in each geographic market, one of which was almost always AT&T. This kept the market small. The RCCs fiercely opposed cellular telephony lest it compete with them. Motorola was their ally, defending its near monopoly in the manufacture of handsets, which remained big, expensive and energy-hungry, with limited bandwidth.

AT&T was banned from making mobile sets under an anti-trust settlement, though its own research arm, Bell Labs, had invented and conceived cellular. But AT&T was sitting on a comfortable monopoly of landlines and saw no need to compete against itself anyway. As late as 1980, when cellular was clearly on the way, AT&T forecast that there might be as many as 900,000 mobile handsets in use in America by the year 2000. In the event there were 109 million. That a telephone company could not see that people wanted to talk to each other is corporate myopia at its most extreme.

In short, the government, in cahoots with crony-capitalist firms with huge vested interests, made the development of cellular service impossible for almost four decades. Who knows what improvements in technology and changes in society it thus prevented? In 1970 the FCC at last suggested allocating some spectrum to cellular, and in 1973 Marty Cooper, vice-president of Motorola, placed the first cellular call with a mobile handset. Yet his own company was lobbying behind the scenes to stop cellular at the very same time, because it had a comfortable monopoly in the (non-cellular) radio communication sets that were allowed. As a result the FCC remained tied up in legal tussles with various litigants for the next decade, and anyway stuck to the false assumption, so common among regulators, that cellular telephony was a 'natural monopoly' anyway, and would therefore have to remain owned by AT&T. Only in 1982, with a new breeze of competition-friendly policy blowing through Washington, did the FCC finally start accepting applications for cellular licences. On 28 July 1984, thirty-nine years to the day after Mr Jett had said it 'won't be difficult' to launch cellular telephony, America's first cellular mobile service went live for the opening ceremony of the Los Angeles Olympics. Who says that the pace of change is breathtaking?

But if America was thus hamstringing itself, why did another continent not steal a march? Some smaller countries did get ahead, but without large markets could not take off, and Europe's regulation of telephony was even more hidebound, being mainly done by nationalized industries themselves with no interest in disrupting their comfortable rent-seeking models. So, on mobile, Europe waited and watched the United States at first. Once it saw the unexpected popularity of 1G (analog) mobile, however, the European Union was stirred into action and set about creating a digital 2G standard called GSM, at the behest of Ericsson, Nokia, Alcatel and Siemens, which owned the key patents. The American firm Qualcomm's rival standard, CDMA, was simply banned in Europe in a clear case of protectionism.

America failed to set a standard till 1995, so in the early 1990s, Europe overtook America in mobile telephony. The FCC's attempt to open up the market got bogged down in political battles. By the late 1980s, GSM had 80 per cent of the world market and Europe's 'industrial policy' seemed to have paid off. But GSM was built for voice, with data as an afterthought, while CDMA was built for data with voice as an add-on. So when 3G arrived around the year 2000, GSM networks could not cope and the world soon switched to CDMA. Not for the last time Europe had shot itself in the foot by closing itself off from global competition.

Thus the story of mobile phones is one in which governments resisted an innovation, in league with vested interests in the private sector. We use smartphones today not because of government regulators, but despite them.

Another slightly less extreme case is the development of drones. Unmanned aerial vehicles, running on batteries, surprised the world in the second decade of the twenty-first century by their sudden ubiquity. Although military, unmanned, radio-controlled aircraft were in widespread use from 2001, the first Wi-Fi-controlled, quadcopter drone aimed at consumers was the French Parrot AR drone, released in 2010. Quadcopter drones quickly found commercial applications in surveying, aerial photography, farming, search and rescue, and other areas. The reaction of governments was to restrict their use with rigid rules that turned out to inhibit innovation by preventing learning. In the United States, as late as 2016, drones were forbidden from flying above 400 feet, out of sight of the operator, near an airport, outside of daylight hours, over people, or if they weighed more than 55 lb. While these may seem sensible precautions, the same effect could have been achieved, so the entrepreneur John Chisholm has argued, by simple rules that said things like: 'drones will operate safely and not harm people or property', then leaving the common law to enforce compliance. Chisholm argues that such an organic rather than imposed regulatory system would be more likely to encourage innovation in how to operate drones and make them good at sensing and avoiding hazards, while providing as much safety. China, with much less restrictive rules, soon came to dominate the industry. American regulations have since been gradually relaxed, but it may be too late.

Likewise, the coming regulation of the digital industry will almost certainly stifle innovation, whatever else it achieves. We know this because the European Union has conveniently carried out an experiment to show this. In 2018 it brought in the General Data Protection Regulation (GDPR), forcing internet content providers to seek consent before using data about people. Though this had some benefits, it also undoubtedly reduced competition within Europe, entrenching the power of big players. Google slightly increased its market share among ad-tech vendors in the three months following the introduction of GDPR, while smaller firms relying on adverts for income saw their market share fall steeply. Many smaller firms outside the EU simply blocked EU content, unable to afford the cost of compliance. American tech firms spent $150bn on GDPR compliance, and Microsoft alone hired an extra 1,600 engineers. But the costs for smaller firms were proportionately greater. Whatever else positive GDPR achieves, it will also have created a barrier to entry preventing innovative smaller firms from challenging the big technology companies. As always, regulation favours incumbents.

When the law stifles innovation: the case of intellectual property

The justification for intellectual property – patents and copyright – is that it is necessary to encourage investment and innovation. With real property rights, people will not usually build a house unless they own the land on which it stands; so they will not invent a drug or write a book unless they can own it. So goes the theory, and, in deference to it, governments, led by the United States, have steadily extended the scope and strength of intellectual property in recent decades. The trouble is, the evidence clearly shows that while intellectual property helps a little, it also hinders, and the net effect is to discourage innovation.

In the case of copyright, the terms of entitlement were extended from fourteen to twenty-eight years in the early twentieth century. In 1976 they were pushed out to the life of the author plus fifty years; in 1998 to life plus seventy years. (So my future great-grandchildren can earn money from this book if it sells well: why?) Copyright was also expanded to include unpublished works while the need to assert it was removed, so it became automatic. The evidence that this has led to an explosion of book writing, or film making, or music making is threadbare. Most people create works of art because they seek influence or fame, as much as money. Shakespeare had no copyright protection, and pirated copies of his plays abounded, but he still wrote. Today, wherever intellectual-property protections are absent or leaky – in the music industry, for example, where 'piracy' has prevailed – there is no diminution in the enthusiasm of creators.

As Brink Lindsey and Steve Teles document in their book *The Captured Economy*, since Napster first made mass file sharing possible in 1999, revenues in the American music industry have fallen steeply, down 75 per cent between 1998 and 2012. Yet the supply of new music albums doubled in twelve years after 1999. Online file sharing in the music industry, after a brief battle, established itself without killing the music industry: performers went back to performing live to make money, rather than sitting back and letting the royalties roll in. Incumbent industries fought every innovation that came along in the world of art: not just music streaming, but video recording of films too.

Meanwhile, in science, you the taxpayer pay for most research, yet the published results are hidden away behind high paywalls in learned journals dominated by three highly profitable firms, Elsevier, Springer and Wiley, whose business model is to sell back to the taxpayer, in the form of library subscriptions, the fruits of his or her investment. Never mind the ethics of this, it massively slows the diffusion of knowledge out of universities to the obvious detriment of innovation.

In 2019 the European Union proposed a directive on online copyright, one part of which would make internet platforms, rather than those who post on them, responsible for determining whether they have permission to post something. A large group of internet pioneers, including Vint Cerf, Tim Berners-Lee and Jimmy Wales, argued that this was a mistake and would hit small startups at the expense of established tech firms: 'Article 13 takes an unprecedented step towards the transformation of the Internet from an open platform for sharing and innovation, into a tool for the automated surveillance and control of its users.'

As for patents, their purpose is to encourage people to innovate, by allowing them a monopoly profit from the patent for a limited period of time, provided they disclose the details of their invention. The analogy with property – explicit in the term 'intellectual property' – is that without a fence around your garden, you will not look after it or improve it. But the analogy is flawed. The whole purpose of new ideas is to share them and allow them to be copied. More than one person can enjoy an idea without exhausting or diminishing it, whereas the same is not true of physical property.

In 2011 the economist Alex Tabarrok argued in his book *Launching the Innovation Renaissance* that the American patent system, far from encouraging innovation, is now discouraging it. Echoing the famous Laffer curve, which shows that beyond a certain point higher tax rates generate less revenue, he drew a graph on a paper napkin to suggest that beyond a certain point stronger patents generate less innovation, because they make it hard to share ideas, and create barriers to entry. The Semiconductor Chip Protection Act of 1984 resulted in more patenting but less innovation in the United States, as semiconductor firms effectively set about stocking 'war chests' of patents to deploy in disputes with each other.

In this book I have told many stories of how patent disputes bogged down innovators in costly disputes with their rivals. Watt, Morse, Marconi, the Wright brothers and others wasted the best years of their lives in court defending their intellectual property. In some cases, they deserve sympathy: after putting in a lot of hard work they saw pirates profit from their ingenuity. But just as often they were pursuing futile vendettas against rivals who deserved at least some credit. In some cases it took government intervention to sort things out. In the years before the First World War, while French aviators made good progress, American ones became bogged down in lawsuits, which put a halt to innovation. A century later, the 'smartphone patent wars' broke out among rival manufacturers, which resulted in a thicket of legal red tape that effectively kept out all but the biggest technology firms.

The argument that some kind of temporary monopoly on the profits from an invention are due to the inventor, in return for publishing the details, seems reasonable. But, except possibly in special cases, such as the drug industry – where years of expensive testing are necessary before the drug can be licensed as saleable – the evidence is weak that this works. For a start, there is no evidence that there is less innovation in areas unprotected by patents. Lindsey and Teles list the various organizational innovations that have happened in companies, unpatented, widely copied and yet enthusiastically invented: the multidivisional corporation, the R&D department, the department store, the chain store, franchising, statistical process control, just-in-time inventory management. Likewise none of the following technologies were patented in any effective way: automatic transmission, power steering, ballpoint pens, cellophane, gyrocompasses, jet engines, magnetic recording, safety razors and zippers. Inventing something gives you a first-mover advantage, which is usually quite enough to get you a substantial reward. Cunning inventors can throw their imitators off the trail with misleading details: Bosch was careful to let Haber reveal only the second-best catalyst recipe for fixing nitrogen.

Another problem is that there is just no evidence from geography and history that patents are helpful, let alone necessary, in encouraging innovation. Take the case of the English clock and instrument makers of the eighteenth century, an industry that was famous for its inventiveness, producing not just high-quality watches and clocks that were envied throughout Europe and became steadily more affordable, but new and precise instruments such as microscopes, thermometers and barometers. The Clockmakers and Spectacle Makers companies maintained what one historian, Christine Macleod, has called 'an inveterate opposition to patents', spending large sums to try to defeat Acts of Parliament that introduced patents. Their argument was that patents 'restricted the free exercise of a skill whose development had always depended on small improvements freely exchanged among craftsmen'.

Neither the Netherlands nor Switzerland had a patent system in the latter part of the nineteenth century, yet both countries were able to nurture innovation. A study by Josh Lerner of 177 cases of strengthened patent policy in sixty countries over more than a century found that 'these policy changes did not spur innovation'. In Japan another study found that the strengthening of patent protection increased neither research spending nor innovation. In Canada a study found that firms which use the patent process intensively were no more likely to innovate.

A further problem is that patents undoubtedly raise the costs of goods. That is the point: to keep competition at bay while the innovator reaps a reward. This slows the development and spread of the innovation. As the economist Joan Robinson put it: 'The justification of the patent system is that by slowing down the diffusion of technical progress it ensures that there will be more progress to diffuse.' But this does not necessarily happen. Indeed, history is replete with examples of bursts of innovation that follow the ending of a patent.

Finally, patents tend to favour inventions rather than innovations: upstream discoveries of principles, rather than downstream adaptation of devices to the market. This has led to the proliferation of what are known as patent thickets: vague intellectual property hedges that block the progress of people trying to move through the intellectual land-scape and develop new products. This is a special problem in biotechnology, where innovators often find themselves infringing patents taken out by others relating to molecules they need to use in just a small part of their work. Start-ups find themselves blocked from following a lead into a new molecular pathway by the unpleasant discovery that another firm has already vaguely patented the use of one of the molecules therein. As Michael Heller argued in his 2010 book *The Gridlock Economy*, this is like a merchant encountering tollbooths all along the route to market: rais-ing prices and suppressing business.

Despite this evidence, industry – the legal industry in particular – has been successful in arguing for much stricter patent protection in recent years. The number of patents issued each year by the US Patent and Trademark Office has quintupled since 1983 to over 300,000 in 2013, at a time when economic growth has slowed: so it does not seem to have helped the economy to grow. Incredibly, a study found that, except in the chemical and pharmaceutical industries, four times as much money is spent litigating over intel-lectual property as reaping rewards from it. Indeed, most of the lawsuits are brought by companies that make no prod-ucts but are simply in the business of buying up patents and litigating those who trespass on them. These are the 'patent trolls', whose activities cost America $29bn in 2011 alone. Blackberry, the Canadian mobile-messaging firm, fell foul of one extremely expensive such troll. Latterly it has turned into a bit of a patent troll itself, suing Facebook and more recently Twitter for infringing what it says are its property rights to such obvious things as mobile messaging, mobile advertising and 'new-message notifications'.

Tabarrok argues for a three-tier patent system, offering two-, ten- or twenty-year patents, with short patents granted much more quickly, easily and cheaply. At the moment, he says that anybody with a novel, non-obvious idea gets a twenty-year patent regardless of whether the innovation cost a billion or twenty dollars. He recognizes that some industries can justify patents better than others. Pharmaceuticals are the most obvious example. If a firm takes ten years and a billion dollars to create, test and prove a drug to be safe and effective, then it seems unfair if others can then pile in with generic copies.

Yet, even in this case, there is an argument to be made against the current patent system. Bill Gurley, a successful tech investor, suggests that pharmaceuticals firms mostly spend their monopoly profits marketing and defending the monopoly itself, rather than seeking new products. The dismal failure of the pharmaceutical industry to find any effective new drugs for diseases like Alzheimer's, or even to sustain its rate of innovation generally, hardly testifies to the effectiveness of the intellectual-property regime. 'You have to imagine where we might be in a world where there are no patents in drugs. I find the idea that no one would work on innovation ridiculous,' Gurley told me.

All in all, the evidence that patents and copyrights are necessary for innovation, let alone good for it, is weak. There is simply no sign of a 'market failure' in innovation waiting to be rectified by intellectual property, while there is ample evidence that patents and copyrights are actively hindering innovation. As Lindsey and Teles put it, the holders of intellectual property are 'a significant drag on innovation and growth, the very opposite of IP law's stated purpose'. They go on:

It is entirely appropriate to strip IP protection of its sheep's clothing and to see it for the wolf it is, a major source of economic stagnation and a tool for unjust enrichment.

When big firms stifle innovation: the case of bagless vacuum cleaners

There is a general tendency for modern Western economies like America's to accumulate barriers to innovation in the form of rent-seeking opportunities, often, though not always, because of regulation. Patents are just one example. The growth of finance is another. Talented people are diverted from more productive occupations into the relatively unproductive, but lucrative, professions of moving money about in speculative ways that are protected from competition by tight regulation and hidden subsidy. The growth of occupational licensing, restricting jobs to those with certain credentials, tends to hinder entrepreneurial disruption. We are effectively reinventing the guilds that often monopolized and stifled commerce in the Middle Ages. In Europe, roughly 5,000 professions are restricted to those with government-mandated licences. In Florida, an interior designer must go to university for four years before being allowed to practise, even if he or she has already qualified as an interior designer in another state. God forbid that some subversive should put the public interest in danger by trying to furnish a Florida apartment in the Alabama style! In Alabama a manicurist must go through 750 hours of training and pass an exam before setting up in business. These barriers to entry are designed to increase the rewards of those already practising.

In 1937 the number of taxis in Paris was capped at 14,000. In 2007 it was capped at 16,000. Did it occur to anyone that consumer interest in taxis might have grown dramatically over this time period? If it did, nobody in government or the entitled companies gave a hoot. It took outsiders like Uber to come along and shake up this complacent industry and offer consumers the benefit of GPS, mobile data and reputational feedback. The resistance to Uber from taxi drivers, say Lindsey and Teles, is a 'powerfully vivid illustration of the conflict between occupational licensing and innovation'. Cities such as Paris and Brussels passed laws to restrict or even ban Uber.

Land-use planning is another brake on innovation. It drives up the price of housing in fast-growing cities, by restricting supply, with the bizarre consequence that people migrate away from innovative areas. Thus, between 1995 and 2000, as the internet was booming, 100,000 more Americans moved out of San Jose, the nerve centre of Silicon Valley, than moved into it – because of housing costs.

Quite how biased towards incumbent technologies the European political system has become was demonstrated by the curious case of the bag-less vacuum cleaner. The British engineer Sir James Dyson invented the Cyclone vacuum cleaner that operates without a bag, so its suction power did not become weaker as it filled with dust, in spite of the engine working just as hard. In September 2014 the European Commission promulgated a set of 'Eco-design and Energy Labelling regulations' whose purpose was to force manufacturers to make more energy-efficient products. Understandably, Dyson's firm was the very first maker of vacuums to support the idea of an energy label informing customers of the energy consumption of the motors in vacuums: its Cyclone product is very efficient especially in the presence of a lot of dust.

The energy label covers the overall energy rating, rated A to G, with A being best and G being worst; annual energy usage in kWh; the amount of dust in the machine's exhaust (A to G); the noise level in decibels; how much dust the machine picks up from carpets (A to G); and how much dust the machine picks up from hard floors and crevices (A to G). However, bizarrely, it then emerged that the European Commission had stipulated that, under these regulations, vacuum cleaners must be tested in the presence of no dust. This disagrees with the International Electro-technical Commission (IEC), an international standards organization, whose standards have been adopted by consumer test bodies and manufacturers all over the world. It also is different from the way other appliances, such as washing machines, ovens and dishwashers are tested, 'loaded', not empty.

Why did the European Commission depart from international practice? The answer emerged via documents revealed under Freedom of Information. The big German manufacturers of vacuum cleaners with bags had been busily lobbying the European Commission. Bagged vacuum cleaners have to increase power usage as they become clogged with dust or they will perform worse. It is a classic case of crony-capitalism: a company lobbying to get the rules written to favour an incumbent technology over an innovative one.

In 2013 Dyson's firm challenged the labelling rules in the EU General Court, arguing that the performance of a vacuum cleaner should be tested in real-world conditions,

and that this might actually include – this may come as
a shock – encountering some genuine dust. It took until
November 2015 for the EU General Court to give its
leisurely decision, and it dismissed Dyson's claims. The
argument it made was that because dust-loaded testing is
not easily 'reproducible', it therefore could not be adopted
in testing. This despite the fact that the international standard does demand the presence of dust. Dyson knew this
to be rubbish, because it always tests its own machines, in
the lab and in actual houses, using real dust, fluff, grit and
debris, including dog biscuits and, bizarrely, two different
kinds of Cheerio cereals (the best innovators are alert to the
variety of human foibles).

Dyson appealed the decision of the General Court to the
European Court of Justice in January 2016. Time passed.
On 11 May 2017 he won. The court said that to reach the
conclusion it had, the General Court 'distorted the facts',
'ignored their own law', 'had ignored Dyson's evidence'
and had 'failed to comply with its duty to give reasons'.
The judges ruled that the test must adopt, where technically
possible, 'a method of calculation which makes it possible
to measure the energy performance of vacuum cleaners in
conditions as close as possible to actual conditions of use'.
Exactly as Dyson had argued. The ECJ then passed the case
back to the General Court to reconsider its verdict, which it
took an absurd eighteen months to do. In November 2018
the General Court ruled at last in favour of Dyson. But by
now Chinese manufacturers were catching up.

The Dyson firm's comment on this expensive and
pointless five-year delay to an innovative technology was
blistering:

The EU label flagrantly discriminated against a specific technology – Dyson's patented Cyclone. This benefited traditional, predominantly German, manufacturers who lobbied senior Commission officials. Some manufacturers have actively exploited the regulation by using low motor power when in the test state, but then using technology to increase motor power automatically when the machine fills with dust – thus appearing more efficient. This defeat software allows them to circumvent the spirit of the regulation.

There is a direct echo here of the diesel scandal. Environmentalists pressed the European Union to promote diesel engines because of their lower carbon dioxide emissions and despite their higher emissions of particulates and nitrogen oxides. German car manufacturers, with their diesel advantage, joined in, only for the 'defeat software' scandal to emerge some years later – computer programs designed to cheat so cars passed emissions tests in the United States.

Regulatory shenanigans cause harm not just by suppressing entrepreneurial energy but also by misdirecting it. The economist William Baumol has argued that if the policy background means that the best way to get rich is by building a new device and selling it, then entrepreneurial energy will flow into innovation, but if it is simpler to profit from lobbying government to set the rules up in favour of an existing technology, then all the entrepreneurial energy will go into lobbying.

It is the European Union's general – if unintentional – hostility to the innovative process that is a likely cause of the recent slow growth of Europe's economies. The odds are stacked too high against the entrepreneur. The EU placed a string of obstacles in the way of digital startups, leaving Europe in the slow lane of the digital revolution, and with no digital giants to rival Google, Facebook or Amazon – unlike China. It installed an extreme version of the precautionary principle in the Lisbon Treaty itself. Both the European Commission and the European Parliament determinedly opposed or hobbled mobile data, vaping, fracking, genetic modification, bagless vacuum cleaners and most recently gene editing, often using dodgy reasoning derived from pressure groups or corporate lobbies for incumbent interests.

In 2016 BusinessEurope produced a long list of examples in which European regulation had affected innovation. The list did include two cases in which regulation stimulated innovation, in waste policies and sustainable mobility. But it contained a much longer list of cases in which EU regulation had hampered change by introducing legal uncertainty, inconsistency with other regulations, technology-prescriptive rules, burdensome packaging requirements, high compliance costs or excessive precaution. It found that the EU Medical Devices Directive resulted in significantly fewer and more expensive new medical devices than would otherwise have come forward, for example. One study found that a medical device takes around twenty-one months to get through the regulatory process in America from application to the regulator to reimbursement, but seventy months in Germany. In the specific case of the Stratos implantable pacemaker, it took fourteen months in America, forty months in France and seventy months in Italy. Frederik Erixon and Björn Weigel argue that Western economies have 'developed a near obsession with precautions that simply cannot be married to a culture of experimentation'.

When investors divert innovation: the case of permissionless bits

Peter Thiel started out as a philosopher, edited the *Stanford Review*, became a lawyer, then started his own venture capital fund. Having founded Paypal, he spotted the potential of Facebook and became one of its earliest investors. Before the 2016 presidential election he did much the same with Donald Trump, cementing his reputation for seeing which way the world was going – and not adding to his popularity in Silicon Valley.

In the mid-2010s, Thiel made the following observation: 'I would say that we lived in a world in which bits were unregulated and atoms were regulated.' Software was evolving through 'permissionless innovation', while physical technology was tied down in regulation that largely stifled change. 'If you are starting a computer-software company, that costs maybe $100,000,' Thiel added. But 'to get a new drug through the [Food and Drug Administration], maybe on the order of a billion dollars or so'. The result was a paucity of startups in drug development.

This is not an argument for wholly deregulating drug discovery, with all the risks of harm to human health. After all, 'Move fast and break things', Facebook's early motto, would be dangerous in medical innovation. Thalidomide is a horrifying reminder of what can happen if drugs are not properly tested – in that case the effect on foetuses was not found out in testing. But it does suggest that regulation may divert investment from innovation in one sector to another and that if government wishes to entice investors into innovation in one area, it must look hard at the regulations that deter experiment.

Permissionless innovation in bits came about partly by accident but quite a bit by design, at least in the United States. Adam Thierer has made the case that a coalition of policymakers from both parties embraced the notion of permissionless innovation as the basis of internet policy starting in the early 1990s. This became the 'secret sauce' that caused the growth of e-commerce. In 1997 the Clinton administration published a 'Framework for Global Electronic Commerce', a remarkably libertarian document. It argued that 'the Internet should develop as a market driven arena not a regulated industry'; that governments should 'avoid undue restrictions on electronic commerce'; that 'parties should be able to enter into legitimate agreements to buy and sell products and services across the Internet with minimal government involvement or intervention'; and that 'where governmental involvement is needed, its aim should be to support and enforce a predictable, minimalist, consistent and simple legal environment for commerce.' This was the approach that stimulated the explosive growth of e-commerce in the following two decades, and it explains why it happened first and foremost in America.

In fact, America went further than that, passing a law that specifically enabled freedom of expression on the internet, through Section 230 of the Telecommunications Act of 1996, which exempted online intermediaries from liability for the content on their sites. It essentially defined them differently from publishers, and of course is the source of today's anxieties about the power and responsibility of the big tech companies such as Facebook and Google. Section 512 of the Digital Millennium Copyright Act of 1998 likewise protected them against copyright infringement.

A key concept in the study of innovation is Baumol's 'cost disease'. This is the economist William Baumol's realization that innovation in one sector can cause an increase in the cost of products or services in another sector if the latter experiences less innovation. If innovation transforms the productivity of labour in manufacturing, then that will drive up salaries throughout the economy, making services more expensive. In 1995, in Germany, a flat-screen television cost about the same as a hip replacement. Fifteen years later, you could get ten flat-screen televisions for the cost of a hip replacement. The salaries of surgeons had increased because of the general increase in the productivity of the economy, but surgeons' own productivity had not increased much, if at all. Thus, allowing innovation only in one sector can be a problem.

Innovation is one of those things that everybody favours in general, and everybody finds a reason to be against in particular cases. Far from being welcomed and encouraged, innovators have to struggle against the vested interests of incumbents, the cautious conservatism of human psychology, the profitability of protest, and the barriers to entry erected by patents, regulations, standards and licences.

12

An innovation famine

We wanted flying cars; instead we got 140 characters.

PETER THIEL

How innovation works

The main ingredient in the secret sauce that leads to innovation is freedom. Freedom to exchange, experiment, imagine, invest and fail; freedom from expropriation or restriction by chiefs, priests and thieves; freedom on the part of consumers to reward the innovations they like and reject the ones they do not. Liberals have argued since at least the eighteenth century that freedom leads to prosperity, but I would argue that they have never persuasively found the mechanism, the drive chain, by which one causes the other. Innovation, the infinite improbability drive, is that drive chain, that missing link.

Innovation is the child of freedom, because it is a free, creative attempt to satisfy freely expressed human desires. Innovative societies are free societies, where people are free to express their wishes and seek the satisfaction of those wishes, and where creative minds are free to experiment to find ways to supply those requests – so long as they do not harm others. I do not mean freedom in some extreme libertarian, lawless sense, just the general idea that if something has not been specifically prohibited, then the assumption should be that it must be allowed: a surprisingly rare phenomenon today in a world where governments try to dictate what you can do as well as what you cannot.

This reliance on freedom explains why innovation cannot easily be planned, because neither human wishes nor the means of their satisfaction are easy to anticipate in the detail required; why innovation none the less seems inevitable in retrospect, because the link between desire and satisfaction is only then manifest; why innovation is a collective and collaborative business, because one mind knows too little about other minds; why innovation is organic because it must be a response to an authentic and free desire, not what somebody in authority thinks we should want; why nobody really knows how to cause innovation, because no one can make people want something.

A bright future

I am no prophet, and I have in any case argued that it is impossible to forecast the progress of innovation. It is almost as easy to be too optimistic about future technology and practice as to be too pessimistic. None the less, there is little doubt that innovation could dramatically change the world in the coming decades. The potential is vast but will only be imperfectly realized because of the gauntloops and ordeals that inventors must run. Here are some wild guesses as to what we could do through innovation over the next generation to improve the lot of human beings and of the other creatures with which we share this planet. This is not the same as what we will allow ourselves to do.

By 2050, when – if alive – I will be ninety-two and probably in need of care, we could be living in a world in which artificial intelligence has transformed care of the old into something much more affordable, humane and efficient. Already today there are telecare devices that monitor the elderly so that their children or their carers can know, without calling in, that they are safe, active and eating. These are proving more popular and effective than panic buttons and less intrusive than endless visits or phone calls. If the result is greater productivity per carer, then more people could afford care and the wages of carers will also rise. I and my generation can hope for an old age of richer entertainment and more tender care than any previous generation.

If some predictions about the potential of treatments for the ageing process itself are right, based on a growing understanding of how to clear out senescent cells from tissues, then the cost of care for the elderly may plummet. By 2050, too, we could have experienced the long-promised 'compression of morbidity' by which people spend a longer time living but a shorter time dying. So far this has eluded us, as we prevent and cure sudden killers like heart disease much faster than we prevent or cure gradual killers like cancer, let alone chronic diseases like dementia. Medical innovation will surely make it possible to live better lives into old age, with senolytic drugs, robotic keyhole surgery, stem-cell treatment and gene-edited cancer treatments, to name but a handful of possibilities. Artificial intelligence could be helping to make medical care cheaper and better, to give doctors and patients back the 'gift of time' in consultations, as Eric Topol calls it, when they really need it.

By 2050, I am convinced, we could have stemmed the rise of allergies and autoimmune disease, largely by recognizing that the cause lies in the lack of parasites, and the lack of diversity of microflora in our guts, to the presence and resistance of which our immune system is adapted. With transplants of microbiota, or supplements of substances that were once supplied by worms and bacteria, we could have banished many autoimmune diseases, perhaps even including autism or other mental conditions. We will have almost certainly banished the problem of anti-microbial resistance with new strategies for keeping one step ahead of lethal bacteria.

By 2050 we could have gained immense improvements in transport. We may not have routine space travel, but we will surely have artificial intelligence to keep us safe on the road and in the air, just as it already helps us navigate. Transport could be much cleaner, too, and the quality of air in our cities could have continued to improve, while the sharing of rides, roads and vehicles could be far more efficient.

By 2050 we could have altered the relationship between government and money, through the use of cryptocurrencies, so as to banish rapid inflation once and for all. We could have used blockchain to cut out some of the middlemen who cost so much: lawyers, accountants and consultants. We could have made crime much harder to do and easier to detect. We could have made tax fairer and government spending less wasteful.

By 2050 'gene drive' – by which, for example, a DNA sequence eliminates one sex in the offspring of the carrier

– could have transformed the practice of wildlife conservation, allowing us humanely to exterminate invasive alien species that threaten others with extinction, or to dial down the population of one species and thus help another, rarer one. Gene editing could have enabled us to bring back the dodo and the mammoth, and gene-edited crops could have made agriculture so productive that it will need far less land, thus enabling us to provide these dodos and mammoths and other species with large new national parks in which to live.

By 2050 we could have replenished the ecosystems of the oceans and repaired the rain forests through innovative policies as well as innovative machines. Growth increasingly means getting more benefits from fewer resources, as exemplified by the dematerialization of the economy.

By 2050 innovation will make it possible to generate sufficient energy to fuel further improbability and prosperity for all with far lower, perhaps even negative, net emissions of carbon dioxide. Probably that will mean a combination of efficient new modular forms of nuclear power, including fusion, as well as a vigorous carbon capture industry in places like the North Sea, combined with greater use of gas and less use of coal, widespread fertilization of plankton in the oceans and further reforestation of the continents.

All this (and much more) is easily within the reach of innovation by the next generation of entrepreneurs. But will we let them do it, or are we strangling the golden goose of innovation?

Not all innovation is speeding up

It is a cliché to say that innovation is speeding up every year. Like a lot of clichés it is wrong. Some innovation is speeding up certainly, but some is slowing down. Take speed itself. In my lifetime of more than sixty years I have seen little or no improvement in the average speed of travel. When I was born in 1958, planes could travel at 600 mph and cars at 70 mph, just as they do today. Congestion on the roads and at airports has made the scheduled travel time between two points often longer today than it once was. A modern airliner, with its high-bypass engines and less-swept wings, is actually designed to go more slowly than a Boeing 707 was in the 1960s – to save fuel. The record for the fastest manned plane, 4,520 mph, was set by the X-15 rocket plane in 1967 – more than half a century ago – and remains unbroken. (The fastest 'air-breathing' plane, the SR-71 Blackbird, set a record of 2,193.2 mph in 1976, which also remains unbroken.) 747s are still flying, fifty years after they were launched. Concorde, the only supersonic passenger plane, is history.

True, there are better roads, more reliable cars (with more cupholders), fewer crashes and so on, so speed isn't everything. But contrast this experience with the change in speed and efficiency of communication and computing, which has been utterly transformed in my lifetime. If cars had improved as fast as computers since 1982, they would get nearly four million miles per gallon, so they could go to the moon and back a hundred times on a single tank of fuel.

The contrast is all the more striking when you look at the science fiction of the 1950s and 1960s, in which transport technology looms large, whereas computers hardly figure at all. The future, we were told, would include routine space travel, supersonic airliners and personal gyrocopters. No mention of the internet, social media or watching movies on mobiles. I recently dug out an old cartoon strip from 1958 about the future called 'Closer than we think'. One image shows a 'rocket mailman', propelled through the air by a personal jetpack, delivering letters to a house.

My grandparents had the opposite experience from what my generation has seen: big changes in transport and few in communication. Born before the motor car or the aeroplane, they lived to see supersonic planes in the sky, wars fought by helicopter and men on the moon. Yet they saw little change in information technology. They were born after the telegraph and the telephone, but died before the mobile phone and the internet. When the last of my grandparents died, a transatlantic phone call to her daughter was still an expensive rarity that usually needed booking through an operator. I suspect that the next half-century will not be nearly so dominated by advances in computing as we are currently expecting, and I have a suspicion that around 2070 there will be essays about the slowdown in changes in information technology, and the acceleration of biotechnology.

The innovation famine

Some people are arguing that we live in an age of innovation crisis: too little, not too much. The Western world, especially since 2009, seems to have forgotten how to expand its economy at any reasonable speed. The rest of the world is making up for this, with Africa in particular beginning to rival the explosive growth rates that Asia achieved in the previous two decades. But most of this is catch-up growth, caused by adopting the innovations already in use in the West.

By contrast, the forces of complacency and stagnation sometimes seem to be winning in Europe, America and Japan. Fredrik Erixon and Björn Weigel, in their book *The Innovation Illusion: How So Little is Created by So Many Working So Hard*, argue that the existential challenge of today's capitalism is to break the habit of both companies' and governments' reluctance to encourage innovation, despite their words.

Schumpeter's 'perennial gale of creative destruction' has been replaced by the gentle breezes of rent-seeking. Corporate managerialism is gradually squeezing the life out of enterprise as big companies in cosy cahoots with big government increasingly dominate the scene. Their bosses shy away from uncertainty, and instead make their companies increasingly bureaucratic. Economists such as Tyler Cowan and Robert Gordon have likewise argued that we are no longer inventing things that really change the world, like toilets and cars, but increasingly playing with trivia like social media.

A symptom of the disease is that companies are sitting on huge cash piles, measured in the trillions, and multinational firms have become net lenders, rather than borrowers, because they cannot see ways to invest their money in innovation. Some big pharmaceutical companies may now make more profit from their financial investments than they do from selling drugs. When big companies do spend money it is often defensively to enforce their patents or protect their market share. Their assets are ageing and they are increasingly apt to play safe. This is partly the fault of diffused ownership, by pension funds and sovereign wealth funds, and the lack of skin in the game that comes with it, which has a tendency to turn entrepreneurs into rentiers, extracting profits from local monopolies achieved through raising barriers to entry via intellectual property, occupational licensing and government subsidy. The dead hand of corporate managerialism then finds that it is easier to control markets than to contest them, to plan rather than experiment. The rapid and continuous growth in the number of 'compliance officers' within firms shows how this plays out. Compliance with regulation almost always hits small companies harder, proportionately, than big ones, thus deterring new entrants with new ideas from entering existing markets. The economist Luigi Zingales argues that most of the time 'the best way to make lots of money is not to come up with brilliant ideas and work hard at implementing them but, instead, to cultivate a government ally.' Of course, many companies still pay lip service to innovation, appointing executives to jobs with the word in the title, and adopting slogans that use the term, but this is often meaningless blather disguising a deep attachment to the status quo.

Globalization, far from challenging this trend, may have entrenched it. Multinationals have absorbed the mentality of planners, rather than entrepreneurs. These factors probably explain the declining dynamism of the American economy, and its rising inequality. The rate of new business formation in the United States fell from 12 per cent a year in the late 1980s to 8 per cent in 2010. The turnover of companies in the main indices has dropped significantly, meaning that incumbents stay in place for longer. Between 1996 and 2014, the proportion of startups begun by people in their twenties halved. Startup rates are falling in sixteen out of eighteen economies, according to an OECD study.

The problem is even worse in Europe, where creative destruction has almost ground to a halt in the cuddly embrace of the European Commission with its tendency to write rules that favour incumbent firms. Of Europe's 100 most valuable companies, none – not one – was formed in the past forty years. In Germany's Dax 30 index, just two companies were founded after 1970; in France's CAC 40 index, one; in Sweden's top fifty, none at all. Europe has spawned not a single digital giant to challenge Google, Facebook or Amazon.

If this line of thought is right, the ability of Western economies to generate innovation has become weaker. To the extent that incomes appear to be stagnating and opportunities for social mobility drying up, the cause is not too much innovation, but too little. 'The troubling reality,' write Erixon and Weigel, 'is that we should fear an innovation famine rather than an innovation feast.' Brink Lindsey and Steve Teles agree: 'The machinery of creative destruction is slowing down, the evidence of which is increasing corporate profits, declining new firm formation, and disturbingly increasing stability of the top firms over time.' But perhaps another part of the world will come to the rescue. Just as Europe seized the innovation baton from an increasingly sclerotic China six centuries ago, perhaps China is about to seize it back again.

China's innovation engine

There is little doubt that the innovation engine has fired up in China. Silicon Valley's will sputter on for a while, but on most metrics California will struggle to attract talent in the future: it is an increasingly expensive, constrained, regulated and taxed place to work. Texas is doing better, and Israel, New Zealand, Singapore, Australia, Canada and even parts of Europe have their bright spots, especially London and its hinterland, but it is likely that in the coming few decades China will innovate on a grander scale and faster than anywhere else. This despite the fact that its politics is authoritarian and intolerant, because a lot of that happens at a level above the entrepreneur, who is surprisingly free of petty bureaucratic rules and delays, so long as he or she does not annoy the Communist Party, and free to experiment. So the lack of political freedom may not matter at first, though it will surely prove a problem in time.

The days when China was a smart copier, catching up with the West by emulating its products and processes, are over. China is leapfrogging into the future. It is wholly mobile in its use of the internet, floating free of fixed computers. In cities at least Chinese consumers no longer use cash, or even credit cards: mobile payments are universal. Digital money, controlled by Tencent and Alibaba, is evolving fast. You mostly no longer find menus in restaurants or cash registers in shops: QR codes are used to pay for and order everything. The cost of mobile data has plummeted there faster than anybody could have imagined. In five years, the price of a gigabyte of data plunged from 240 renminbi to just one.

Firms like WeChat started out as social media companies but are now providing everything consumers want: mobile wallets, apps for ordering taxis or meals, means of paying utility bills and much more. Things that require five different apps in the West can be done on one app in China. Companies like Ant Financial are reinventing financial services, with 600 million users managing not just their money but their insurance and other services, all through a single app.

As for discovery and invention, China is just as innovative, plunging into artificial intelligence, gene editing, nuclear and solar energy, with a gusto that the West can only dream of. The pace is breathtaking: 7,000 miles of new motorways a year over the past decade; train lines and metro networks that would take decades in the West appearing in a year or two; data networks bigger, faster and more comprehensive than anywhere else. This infrastructure spend is not innovation, but it surely helps it happen.

What explains this speed and breadth of innovation fury? In a word, work. Chinese entrepreneurs are dedicated to the 9–9–6 week: 9 a.m. to 9 p.m., six days a week. That was what Americans were like too when they changed the world (Edison demanded inhuman hours from his employees); and Germans when they were among the most innovative people; and Britons in the nineteenth century; and Dutch and Italians before that. Willingness to put in the hours, to experiment and play, to try new things, to take risks – these characteristics for some reason are found in young, newly prosperous societies and no longer in old, tired ones.

The West may still do clever new things in finance, science, the arts and philanthropy, but it is slowing down in innovating the products and processes that affect everyday life. Bureaucracy and superstition get in the way of anybody who tries. London takes three decades to (not yet) build a single new runway for its main airport, while the consultants get rich from investigating what will happen to every newt, bat and noise meter within miles. Brussels cogitates for years on whether it is a good idea for somebody even to try to make a crop resistant to insects. Washington lays on a feast for regulators, lawyers, consultants and rent seekers sucking the vital juices from entrepreneurial enterprises. Central banks look down their noses at cryptocurrencies and digital fin-tech. Just as in Ming China, Abbasid Arabia, Byzantium and Ashokan India before them, these mature civilizations lose the innovation bug and pass the buck.

Regaining momentum

Yet if the world is to rely on China to do its innovation, it will become an uncomfortable place. Chinese citizens are subject to arbitrary and authoritarian restraints that the West long ago shook off. Democracy does not exist and free speech is impossible. I repeat: the stories of innovation that I have documented in this book teach a lesson that it relies heavily on freedom. Innovation happens when ideas can meet and mate, when experiment is encouraged, when people and goods can move freely and when money can flow rapidly towards fresh concepts, when those who invest can be sure their rewards will not be stolen.

The West may be slowly forgetting to allow this to happen through bureaucratic strangulation, but China will surely stifle it through political authoritarianism. In an authoritarian system it will be all too easy for incumbent businesses, even those that started out as plucky outsiders, to raise barriers to entry against innovation. Who then will pick up the challenge?

Perhaps India – a vast country now reaching middle-income living standards, with a well-educated population and a long tradition of free enterprise and spontaneous order. India's innovation is accelerating noticeably, with technologies like the use of biometric identification, using fingerprints and irises, for welfare payments and banking already showing signs of leapfrogging both the West and China. India's drug industry is rapidly moving from generics to innovative medicines.

Or perhaps Brazil, a country that has increased its patent applications by 80 per cent in just ten years. The country has an enviable cluster of expertise in fin-tech, ag-tech and apps. Embraco, the world's largest maker of compressors, is looking to transform refrigerators so they have no need of compressors at all.

I hope somebody does keep innovation happening, because without innovation we face a bleak prospect of stagnant living standards leading to political division and cultural disenchantment. With it, we face a bright future of longevity and health, more people leading more-fulfilled lives, astonishing technological achievements and a lighter impact on the planet's ecology.

Of all the lessons taught by the stories told in this book, I think the most relevant is Thomas Edison's. He was only one of many people who conceived the idea of the light bulb, but he was the one who turned it into a practical reality. He did so not by genius, but by experiment. As he told several interviewers, genius is 1 per cent inspiration and 99 per cent perspiration (he sometimes said 2 per cent and 98 per cent). 'I tell you genius is hard work,' he added, 'stick-to-it-iveness, and common sense.' I repeat that Edison tested 6,000 plant materials till he found the right kind of bamboo for the filament of a light bulb. The perspiration, not the inspiration, is the bit that much of the West has forgotten or forbidden. It is the impossibility of repeated experiment that prevents nuclear power becoming safer and cheaper, that prevented Golden Rice saving lives sooner than it has, that slows down the development of new medical treatments. And it was a bounty of repeated experimentation that led to the growth of the internet and the expansion of the world of digital communication. Somehow we must find a way to reform the regulatory state so that while keeping us safe it does not prevent the simple process of trial and error on which all innovation depends.

Innovation is the child of freedom and the parent of prosperity. It is on balance a very good thing. We abandon it at our peril. The peculiar fact that one species above all others has somehow got into the habit of rearranging the atoms and electrons of the world in such a way as to create new and thermodynamically improbable structures and ideas that are of practical use to the wellbeing of that species never ceases to amaze me. That many members of the species show little curiosity about how this rearranging comes about, and why it matters, puzzles me. That many people think more about how to constrain rather than encourage it worries me. That there is no practical limit to the ways in which the species could rearrange the atoms and electrons of the world into improbable structures in the centuries and millennia that lie ahead excites me. The future is thrilling and it is the improbability drive of innovation that will take us there.

Afterword

A virus reminds us of the value of innovation

I finished the final draft of this book in November 2019, before the coronavirus that caused the Covid-19 pandemic burst upon the world. A few stories about a new kind of viral 'pneumonia' were percolating out of China in December and January, but the World Health Organization was still insisting as late as 14 January that there was no evidence for human-to-human transmission, and that it did not seem like a big threat. In February, it became clear that the epidemic was out of control. Publication of my book in Britain was then delayed by the lockdown that came into force on 23 March 2020. This has given me time to write an afterword about the implications of the disease for the argument of this book.

My book is not specifically about medicine, but it includes a lot of examples of innovation drawn from the world of infectious disease and public health: the stories of vaccines against smallpox, polio, rabies and whooping cough, the story of penicillin, of insecticide-treated bednets preventing malaria, of oral rehydration therapy curing cholera, of the water closet and of the chlorination of water supplies. All of them felt, while I was writing them, safely in the past: spectacular victories against contagious disease, an enemy that was no longer capable of killing large numbers of people in the richer countries of the world and was in rapid retreat even in the poorer countries. But all of them also show the incomparable value of innovation. And that will be just as true with this one: an innovation will beat it, whether that be a vaccine, a pill, a means of enforcing quarantine or a social-media app to track and trace contacts with infected people.

I confess I did not think a pandemic this bad was likely. After all, time and time again warnings of pandemics came to nothing. Ebola was going to kill us by the million in 1995. BSE (mad-cow disease) in 1996. SARS in 2003. Bird flu in 2006. Swine flu in 2009. Bird flu again in 2013. Ebola again in 2014. Zika in 2016. In every case there were apocalyptic predictions from mathematical models, articulated by experts and relayed by the media, about hundreds of thousands or even tens of millions dying. 'Wolf!' went the cry each time. In each case, people died, but in their hundreds or thousands, not hundreds of thousands. No wolf. We grew tired of being told that the latest version of influenza was going to be as bad as 1918 only to find it was nothing of the kind. Stockpiles of Tamiflu, the possibly helpful medication for influenza, sat unused in warehouses after pharmaceutical companies lobbied governments to buy it in bulk in 2009. HIV was the only new virus caught from other animal species that proved capable of causing a global pandemic in my lifetime. It has killed around 20 million people but is now mercifully curable and declining as a cause of death.

In Aesop's fable, the boy who cried 'Wolf!' was not believed when the wolf at last showed up, because he had given too many false alarms. Pandemics that did not happen were not the only false alarms. I've watched exaggerated claims come and go about the population explosion, peak oil, nuclear winter, acid rain, the ozone hole, pesticides, species extinction rates, genetically modified crops, sperm counts, ocean acidification and especially the millennium bug. Note: these are real issues, but subject to wild hyperbole in the media. Competitive auctions of unjustified alarm are meat and drink to journalists and politicians. 'The whole aim of practical politics,' said H. L. Mencken, 'is to keep the populace alarmed (and hence clamorous to be led to safety) by an endless series of hobgoblins, most of them imaginary.' Nor is the explanation that the panics caused the world to seek out solutions. The millennium bug was a damp squib in those countries and businesses that did nothing as well as those that prepared.

In the weeks leading up to the 2020 Covid pandemic the Western media was full of apocalyptic warnings, wild protests and angst-ridden pleas about the greatest crisis ever to face humankind – the prospect that the average temperature of the air might rise by a few tenths of a degree per decade, mostly at night, in the north and in winter, until eventually this had net negative impacts. Although climate change is a serious, long-term problem to be tackled, the claims of imminent catastrophe and the deaths of millions, even billions, had been escalating for more than three decades at this point, while in fact deaths from famine, storm, flood and drought had been falling steeply throughout the period. Such exaggerated claims were not confined to the extreme fringe of protest movements like Extinction Rebellion. The World

Health Organization, an organisation specifically charged with watching out for pandemics, announced in 2015 that 'climate change is the greatest threat to global health in the 21st century' – greater than pandemics, in other words.

Then a real wolf showed up. Dulled into complacency, we did not believe it. At first, I was among those who thought this coronavirus would prove to be just another kind of 'bad cold' or would quickly evolve into a flu-like fever of no great virulence. I was reassured by medical friends that this was the case – and yes, I had been lulled by the failure of previously predicted pandemics to materialise. Yet I certainly knew a contagious pandemic was possible. In 2010 in *The Rational Optimist* I wrote that despite all the good things that will happen in the current century, setbacks are likely and nothing was guaranteed. I added: 'pandemic flu may yet make the twenty-first century a dreadful place'. Back in 1999 in an essay on the future of disease, I wrote that if a new pandemic did happen it would be a virus, not a bacterium, fungus or animal parasite, and that we would catch it from a wild animal, because we have already caught the diseases of our domesticated animals (measles probably came from cattle). 'My money is on bats,' I wrote. Bats, of which there are over a thousand different species, are highly gregarious, and therefore ideal hosts for respiratory viruses. Like us they gather in large crowds; like us they fly long distances; like us they continually cough out noises – to navigate as well as communicate. If we persisted in creating conditions in which viruses can be easily transmitted and amplified, then we would persist in experiencing waves of new viral epidemics.

We now know (thanks to innovation in gene sequencing) that the virus that causes Covid-19 is one of many coronaviruses found in bats and that it already had the capacity to unlock human cells through a particular receptor known as ACE2 even before it jumped species. We do not yet know if the virus reached us through amplification in an unhygienic 'wet market' where live wildlife of many different species is mixed up and put on sale for food and medicine, or if it might possibly have escaped from a laboratory experiment, or a bit of both. Such accidents do happen and the Hubei Center for Disease Control and Prevention laboratory, where scientists such as Tian Jun-Hua led virus-hunting expeditions to local bat caves, is just a few hundred yards from the wet market in Wuhan where the epidemic seems to have begun.

If you knew where to look in the scientific literature, the warnings were crystal clear. 'The presence of a large reservoir of SARS-CoV-like viruses in horseshoe bats, together with the culture of eating exotic mammals in southern China, is a time bomb,' wrote four Hong Kong scientists in 2007. 'Bat–animal and bat–human interactions, such as the presence of live bats in wildlife wet markets and restaurants in southern China, are important for interspecies transmission of [coronaviruses] and may lead to devastating global outbreaks,' said some of the same scientists in 2019.

The neglect of innovation in vaccines and diagnostics

Yet such warnings were ignored, wildlife wet markets continued to flourish and the world was not prepared. Worse than that, the world had been neglecting innovation in exactly the areas we most needed. Vaccine development, for example, had languished in the twenty-first century as an orphan technology, insufficiently encouraged by governments and the World Health Organization, which preferred to use their public-health budgets to lecture their subjects on diet or climate change. Ignored, too, by the private sector because new vaccines are not profitable things to make. By the time you have developed one for a new epidemic, the epidemic may be over, and if it is not, you will be under enormous pressure to give the vaccine away for free during an emergency. Plus, if the thing works, it is only needed once per person and soon does itself out of business, unlike, say, a statin. In the case of the West African ebola epidemic of 2014, an experimental vaccine was first developed in November, but within seven months the epidemic was over and the firm was struggling even to find volunteers for the latest trials. Sympathy for the plight of pharmaceutical firms in such a case is unlikely, so many have decided to steer clear of vaccines.

In Chapter 2, I described the extraordinary achievement of Pearl Kendrick and Grace Eldering in developing a vaccine for whooping cough in just four years in the 1930s, saving countless lives with their ingenuity and hard work, done partly in their spare time. It is a shock to realise that it still takes years to make a vaccine despite the fact that – unlike those two women – we now know what genes are made of, how the immune system works, how proteins are synthesised, what the genetic codebook is and much, much more. Innovative ideas for making vaccines, such as messenger-RNA vaccines, where the jab effectively puts the instructions for making the vaccine into the body, may come to our rescue this time, but they have been developed by small teams in a few laboratories with limited funds. Vaccine development, warned Wayne Koff, president of the

Human Vaccines Project, in 2019 before the pandemic, 'is an expensive, slow and laborious process, costing billions of dollars, taking decades, with less than a 10 per cent rate of success . . . There is clearly an urgent need to determine ways to improve not just the effectiveness of the vaccines themselves but also the very processes by which they are developed.' In short, innovation was and is badly needed.

It was precisely to solve this problem that the Coalition for Epidemic Preparedness Innovations (CEPI) was founded in 2017, with money from the Wellcome Trust, the Gates Foundation, and the Indian and Norwegian governments. But somebody – probably the World Health Organization – should have done this far sooner. By 2020, CEPI had barely started on the road to inventing a vaccine-development platform that is adaptable to any new disease and ready to go. By far the biggest failure of the 2020 pandemic is the failure to have done enough innovation in the field of vaccines. It took me by surprise, for sure.

This failure applies to other things than vaccines. As the pandemic spread, countries soon realised that their chief weapon against the virus was diagnostic testing to identify and isolate people with the disease. Some countries like South Korea and Germany quickly enlisted the private sector, farming out the development, manufacture and distribution of tests to firms and enabling the process of trial and error by which the private sector discovers effective and efficient solutions. Others, like America and Britain, tried to keep a government monopoly on the testing, using the excuse that this was the only way to achieve quality control. The Centers for Disease Control and Prevention in Atlanta at first 'sought to monopolise testing, discouraged the private sector developing its own tests and misled state and local authorities about efficacy of its tests', according to one study by Matthew Lesh at the Adam Smith Institute. But after heavy criticism, the US government changed policy, decentralised the system and the private sector rapidly expanded testing to hundreds of thousands of tests a day.

Britain continued to send all samples to its own laboratories and chose 'to develop and encourage the use of its own diagnostic tools, rather than seeking the development of a range of private sector tools and providing fast-track approval', according to the same report from the Adam Smith Institute. By the middle of March, with case numbers shooting up far beyond the ability of the state laboratories to cope, Britain just gave up testing people with symptoms unless they entered hospitals, rather than outsource the testing – a truly bizarre obsession with putting centralised command and control before patients.

The antipathy within Britain's National Health Service to innovative products from outside the system was a problem that had already been identified a few years before by the government's adviser on life sciences, Sir John Bell. The size of the in-vitro diagnostics market in Britain per head of population was less than half that of Germany. According to the British In Vitro Diagnostics Association, 'the NHS is too inflexible when it comes to adopting new IVD tests. Typically, solutions are still thought of as pharmaceuticals and consideration is not given to how IVDs could be adopted in the system to improve outcomes.' One new test that rapidly separates the 20 per cent of patients with chest pain in Accident and Emergency wards who need treatment from the 80 per cent who can be safely sent home, freeing up beds and saving money, was sold by a British firm all over the world but not in Britain. In 2020 this reluctance to use tests was now costing lives.

As countries grappled with the pandemic, so they encountered obstacles to progress that had seemed reasonable in 'peacetime' but now looked ridiculous. Patents once held by Theranos (see Chapter 10) threatened to block the development of diagnostic tests at one point, till the new holders backed down. Many regulations simply got in the way of speed without adding to safety. Rules intended to improve the safety of medical devices turned out to have caused immense delays, undoubtedly deterring innovators from even trying to develop new diagnostic tests. Remember that while there will be some evidence of the people who tried and failed, there will be little evidence of the people who did not even try to start businesses because it was too difficult to bring an innovation to market. As I said in Chapter 11, one pacemaker took seventy months to get licensed for use in Italy.

This is also true of therapies. Whereas the battle against bacteria has been won partly with chemical weapons called antibiotics (notwithstanding the growing problem of resistance), the progress of inventing effective antiviral drugs has been much slower and less successful. The few antivirals that do exist are mostly highly specific to particular viruses. 'The paucity of true broad-spectrum antiviral agents leaves a major chasm in preparedness for viral infectious disease emergencies,' concluded one review presciently, months before the pandemic began. This is mainly because viruses have no biochemistry of their own – they merely subvert that of their host to their own needs – so there are few targets to attack without hurting the host. Another reason is that, as with vaccines, drugs to stop pandemics may be needed only once a decade or less, and so cannot turn into money spinners for pharmaceutical firms that invest huge sums in proving their worth in clinical trials.

But as the example of HIV shows, antivirals are possible, in that case mainly protease inhibitors, which stop the virus using a specific host enzyme to enter cells. Who knows how many such antiviral drugs would have been developed if the world had been spending more on the problem? As it was, protease inhibitors and monoclonal antibodies developed for use against HIV and RNA polymerase inhibitors developed for use against ebola both gave drug firms somewhere to start on cures for Covid. In a fine example of serendipity, Fujifilm's antiviral product, favipiravir (sold as Avigan), is one of the few antivirals showing promise against more than one kind of virus. Fujifilm avoided Kodak's fate by diversifying into other chemical and medical businesses in the early 2000s, and in 2008 it acquired Toyama Chemical, which had one promising drug candidate, favipiravir, developed by a virologist, Kimiyasu Shiraki, in the search for a herpes cure, but now showing promise against influenza. In 2014 the drug was tested in ebola patients in Guinea, with only slightly promising results. Initial trials on coronavirus patients in China were promising, so Fujifilm accelerated production massively in the hope that it might have a cure. Here's a good example of how innovation works: the battle against one virus that led to another to another and finally to being there when needed for a global pandemic, and in the hands of a camera company of all things.

By the time you are reading this, an effective antiviral cure for Covid-19 may have emerged from the crowd. This may even be the first pandemic to be halted with antiviral therapy rather than a vaccine. There is no way of knowing, at this moment, which horse to back. One review listed thirty potential runners: as always with innovation, trial and error will decide, but the chances that none of them can cure or ameliorate this disease are small. An effective antibody test may have become available to identify those who have had the disease as well as those who have it. One way or another we will eventually escape from this nightmare and resume economic activity. We may even count a few blessings as we do so that this was not a worse brush with disaster. For instance, although Covid-19 was very dangerous for old people with underlying health issues, it proved almost completely harmless to children, quite unlike flu, plague, smallpox and most of the other diseases of the past which killed the young and the very old with equal efficiency.

Digital innovation alleviates the isolation of quarantine

One immediate effect of the virus was to shut down the world economy. As it spread through Europe and other continents in March 2020, governments took the tough decision to lock down their populations, telling all but essential workers to stay at home. The impact has been devastating, but it is worth reflecting on how much worse it would have been twenty years before, when video conference calls to grandchildren were not an option for most people, online meetings were impossible and online shopping barely existed. The existence of broadband made the lockdown far more productive for some than it would have been before, and probably caused many people to rethink their commuting habits.

In this way the pandemic should surely unleash a torrent of innovation. It looks like it has tipped a critical mass of the population into video-conferencing for the first time, introducing many of us to systems like Zoom, Teams, Facetime and Skype. Maybe it is naïve, but I hope I can attend literary festivals all over the world to publicise this book without slogging through airports, security checks, hotel lobbies and time zones. Surely, too, working from home will be much more common and less stigmatised after the pandemic fades, as will virtual visits to the doctor and other professionals. Already in some countries workers are demanding more rights to work from home. Productivity in the healthcare, accounting and legal sectors can surely begin to accelerate, instead of stagnating, if tele-meeting innovations are permitted to help here. There has been a surge of free data sharing to help researchers, too, and the opening up of access to scientific papers. The profiteering oligopolies of scientific publishing will surely not be the same after this. Lack of free access to the research that we all pay for has long been a scandal.

The demise of cash has probably been accelerated, as has the decline of high-street retailing in favour of online. Some social events like video-conferenced dinners – invented during the lockdown for friends to see each other – will fade away again as we go back to real parties, but not the family chats among those living far from parents or grandparents.

Digital innovation has medical uses, too. The ability to trace the movements of people by their smartphones, and thus track down contacts with infected people, was widely used in countries such as South Korea and will be key to managing the disease everywhere. This technology is already being made even more effective, safer and more confidential. For example, smartphones will exchange encrypted but meaningless messages that leave a trail of anonymised data revealing when two people came close to each other, one of whom was infected. Thus you could be warned that you should self-isolate by your phone without anybody in government or a big tech firm knowing your name or your habits. Contact tracing need not lead us into the clammy embrace of Big Brother.

It's not saying no that is the problem, it's saying yes slowly

These few bright spots notwithstanding, there will be terrible economic damage to be repaired when the pandemic ends. A deep recession is inevitable. Unemployment will greatly increase, inflation will rocket, many people's debts will become unsustainable, trade protectionism will spread. These shocks will undoubtedly hit the poor the hardest, ruining many lives. And it is here that the world must learn the main lesson of this book. Prosperity comes from innovation, innovation comes from the freedom to experiment and try new things, and freedom relies upon sensible regulation that is permissive, encouraging and quick to give decisions. By far the surest way to rediscover rapid economic growth and help the poorest will be to study the regulatory delays and hurdles that were swept aside to encourage innovators in medical devices and therapies during the pandemic, then see whether such reforms could be made permanent and applied to other parts of the economy, too.

Again and again during the crisis I have spoken to entrepreneurs and scientists frustrated at the unnecessary delays introduced by bureaucratic procedures. Ten days to finalise a contract for the government to purchase diagnostic tests, for example. The lack of urgency displayed by administrators, consultants and legal negotiators, especially in the public sector, is a problem in a crisis but it should have been a problem all along. Whether it is approving a new medical device or building a new airport runway, the process of decision-making has become lethargic to the point of paralysis and encrusted with the requirement to reward legions of unelected 'consultees' for their leisurely attention. The problem for entrepreneurs, watching their capital dwindle as they try to bring an innovation to market, is not so much that regulators say no, but that they take an age to say yes. If we are to return to prosperity after the virus, that has to change.

Politicians should go further and rethink their incentives for innovation more generally so that we are never again caught out with too little innovation having happened in a crucial field of human endeavour. One option is to expand the use of prizes, to replace reliance on grants and patents. The famous Longitude Prize was a £20,000 prize offered in 1714 for the first person to solve the problem of accurately measuring one's longitude at sea within thirty minutes, announced after the failure of expert navigators and astronomers to crack the problem. It eventually elicited a solution from such an unexpected direction – accurate and robust clocks made by a humble clockmaker – that the authorities were reluctant to grant it for many years, to John Harrison's fury. A modern Longitude Prize, offering £8m to a point-of-care diagnostic device to prevent the over-prescribing of antibiotics, was set up in 2014 and remains unclaimed.

Similar serendipity happens today. One study of the online problem-sharing forum known as Innocentive, where individuals, companies and organisations can post details of problems that are baffling them and offer rewards for crowd-sourced solutions, found that 'the further the focal problem was from the solvers' field of expertise, the more likely they were to solve it'. Just like John Harrison. Innocentive has attracted 400,000 contributors from 190 countries and awarded more than $20m to successful solutions.

In March 2020, the economist Tyler Cowen announced a series of modest prizes to reward innovations in social distancing, online worship, easier ways to work from home and treatments for Covid-19. Prizes, he says, are ideal 'when you don't know who is likely to make the breakthrough, you value the final output more than the process, there is an urgency to solutions (talent development is too slow), success is relatively easy to define, and efforts and investments are likely to be undercompensated'.

Yet such a list applies to almost all fields of human endeavour, not just this pandemic. Why don't we do more with prizes? The Nobel-winning economist Michael Kremer came up with a concept that would fine-tune prizes as incentives to innovate, called the Advance Market Commitment. After all, there is no point in giving a prize to a company that invents a vaccine if the firm pockets the reward but then decides not to produce the vaccine because it cannot recoup the costs of doing so. In 2007 the Gates Foundation committed $1.5bn to a prize fund to find a vaccine for pneumococcus bacteria for use in developing countries. Such a vaccine would be of most use to people who could not afford to pay for it, so no pharmaceutical company could make money from inventing it, however long the patent. But rather than a single lump sum prize, the companies were invited to bid for ten-year contracts to develop and manufacture the vaccine. The prize money effectively topped up the sum received by the pharmaceutical firm for every vaccine sold. The result of the auction was three good vaccines, costing $2 per dose, which have been given to 150 million children, saving 700,000 lives.

Governments could also consider buying out patents to free up innovation. Anton Howes of the Royal Society of Arts in London argues that such buyouts have worked well in the past. The French government bought out Louis Daguerre's patent for photography in 1839 and made it freely available to all, unleashing a burst of creative innovation. The recent expiry of 3D printing patents has led to a burst of innovative activity, which could have come a decade sooner if the patents had been bought out. In 1998 Kremer came up with a way of valuing patents so they can be bought out at the right price, using auctions: a bunch of different patents are auctioned for private sale, without the bidders knowing which the government intends to buy. The government steps in at the price discovered by the auction. If it did so rarely, then the private bidders need not be unduly discouraged from taking part. Howes argues: 'As we look to fight coronavirus and any future pandemics, we should perhaps consider which patents – for antivirals, vaccines, ventilators, and other hygienic equipment – might be bought out in order to remove . . . innovation bottlenecks.'

I ended my book in a mood of unaccustomed pessimism, lamenting the innovation famine that seems to have gradually developed, courtesy of complacent big companies, bureaucratic big governments and neophobic big protest groups. With some exceptions, mostly in the digital world, the innovation engine is sputtering and society is not seeing as many new products and services of value as it needs. Covid-19 has driven home that message in emphatic fashion. It is time to let innovation work.

PL Digital Hub
PH. 01 4678909

SOURCES AND FURTHER READING

Introduction: The Infinite Improbability Drive

Adams, Douglas. *The Hitchhiker's Guide to the Galaxy*. Pan Books, 1979.

Christiansen, Clayton. *The Innovator's Dilemma*. Harvard Business Review Press, 1997.

Hogan, Susan. '"Home of sliced bread": a small Missouri town champions its greatest thing'. *Washington Post*, 21 February 2018.

Maddison, Angus. *Phases of Capitalist Development*. Oxford University Press, 1982.

McCloskey, Deirdre. *The Bourgeois Virtues: Ethics for an Age of Commerce*. University of Chicago Press, 2006.

McCloskey, Deirdre. 'The great enrichment was built on ideas, not capital'. Foundation for Economic Education, November 2017.

Mokyr, Joel. *The Gifts of Athena: Historical Origins of the Knowledge Economy*. Princeton University Press, 2002.

Mokyr, Joel. *A Culture of Growth: The Origins of the Modern Economy*. Princeton University Press, 2016.

Petty, William. *Treatise on Taxes and Contributions* (pp. 113–14), 1662.

Phelps, Edmund. *Mass Flourishing: How Grassroots Innovation Created Jobs, Challenge and Change*. Princeton University Press, 2013.

Strauss, E. *Sir William Petty: Portrait of a Genius*. The Bodley
   Head, 1954.

1. Energy
Bailey, Ronald. 'Environmentalists were for fracking before they
   were against it'. *Reason Magazine*, 5 October 2011.
Boldrin, Michele, David K. Levine and Alessandro Nuovolari.
   'Do patents encourage or hinder innovation? The case of the
   steam engine'. *The Freeman: Ideas on Liberty*, December
   2008, pp. 14–17.
Cohen, Bernard. *The Nuclear Energy Option*. Springer, 1990.
Constable, John. 'Energy, entropy and the theory of wealth'.
   Northumberland and Newcastle Society, 11 February 2016.
Friedel, Robert and Paul Israel. *Edison's Electric Light*. Johns
   Hopkins University Press, 1986.
Jevons, William Stanley. *The Coal Question: An Inquiry
   Concerning the Progress of the Nation and the Probable
   Exhaustion of Our Coal-Mines*. Macmillan, 1865.
McCullough, David. *The Wright Brothers*. Simon and Schuster,
   2015.
Rolt, L. T. C. *Thomas Newcomen: The Prehistory of the Steam
   Engine*. David and Charles/Mcdonald, 1963.
Selgin, George and John Turner. 'Watt, again? Boldrin and
   Levine still exaggerate the adverse effect of patents on the
   progress of steam power'. *Review of Law and Economics 5*,
   2009: 7–25.
Smiles, Samuel. *Story of the Life of George Stephenson*. John
   Murray, 1857.
Smith, Ken. *Turbinia: The Story of Charles Parsons and His
   Ocean Greyhound*. Tyne Bridge Publishing, 2009.
Swallow, John. *Atmospheric Engines*. Lulu Enterprises, 2013.
Swan, Kenneth R. *Sir Joseph Swan*. Longmans, 1946.
Triewald, Marten. *Short Description of the Atmospheric Engine*.
   1728. (Published in English by W. Heffer and Sons, 1928.)
Vanek Smith, Stacey. 'How an engineer's desperate experiment
   created fracking'. National Public Radio, 27 September 2016.

Weijers, Leen, Chris Wright, Mike Mayerhofer, et al. 'Trends in the North American frac industry: invention through the shale revolution'. Presentation at the Society Petroleum Engineers Hydraulic Fracturing Technology Conference, 5–7 February 2019.

Zuckerman, Gregory. 'Breakthrough: the accidental discovery that revolutionized American energy'. *Atlantic*, 6 November 2013.

2. Public health

Bookchin, Debbie and Jim Schumacher. *The Virus and the Vaccine: Contaminated Vaccines, Deadly Cancers, and Government Neglect*. St. Martin's Press, 2004.

Brown, Kevin. *Penicillin Man: Alexander Fleming and the Antibiotic Revolution*. The History Press, 2005.

Carrell, Jennifer Lee. *The Speckled Monster: A Historical Tale of Battling Smallpox*. Penguin Books, 2003.

Darriet, F., V. Robert, N. Tho Vien and P. Carnevale. *Evaluation of the Efficacy of Permethrin-Impregnated Intact and Perforated Mosquito Nets against Vectors of Malaria*. World Health Organization Report, 1984

Epstein, Paul. 'Is global warming harmful to health?' *Scientific American*, August 2000.

Gething, Peter et al. 'Climate change and the global malaria recession'. *Nature* 465, 2010: 342–5.

Halpern, David. *Inside the Nudge Unit*. W. H. Allen, 2015.

MacFarlane, Gwyn. *Alexander Fleming: The Man and the Myth*. Harvard University Press, 1984.

McGuire, Michael J. *The Chlorine Revolution: Water Disinfection and the Fight to Save Lives*. American Water Works Association, 2013.

Reiter, Paul. The IPCC and Technical Information. Example: Impacts on Human Health. Evidence to the House of Lords Select Committee on Economic Affairs, 2005.

Ridley, Matt. 'Britain's vaping revolution: why this healthier alternative to smoking is under threat'. *Sunday Times*, 8 July 2018.

Shapiro-Shapin, Carolyn. 'Pearl Kendrick, Grace Eldering, and the pertussis vaccine'. *Emerging Infectious Disease* 16, 2010: 1273–8.
Wortley-Montagu, Lady Mary. [1994.] The Turkish Embassy Letters. Virago.

3. Transport
Davies, Hunter. *George Stephenson: The Remarkable Life of the Founder of the Railways*. Sutton Publishing, 1975.
Grace's Guide to British Industrial History. 'Bedlington ironworks'. https://www.gracesguide.co.uk/Bedlington_Ironworks.
Harris, Don. 'Improving aircraft safety'. *The Psychologist* 27, 2014: 90–95.
Khan, Jibran. 'Herb Kelleher's Southwest Airlines showed the value of playing fair'. *National Review*, 10 January 2019.
McCullough, David. *The Wright Brothers*. Simon and Schuster, 2015.
Nahum, Andrew. *Frank Whittle: The Invention of the Jet*. Icon Books, 2004.
Parissien, Steven. *The Life of the Automobile: A New History of the Motor Car*. Atlantic Books, 2014.
Smil, Vaclav. *Prime Movers of Globalization: The History and Impact of Diesel Engines and Gas Turbines*. MIT Press, 2013.
Smiles, Samuel. *The Life of George Stephenson and His Son Robert*. John Murray, 1857.
Smith, Edgar. *A Short history of Naval and Marine Engineering*. Cambridge University Press, 1938.
Wolmar, Christian. *Fire and Steam: How the Railways Transformed Britain*. Atlantic Books, 2007.

4. Food
Balmford, Andrew et al. 'The environmental costs and benefits of high-yield farming'. *Nature Sustainability* 1, 2018: 477–85.
Brooks, Graham. 'UK plant genetics: a regulatory environment to maximise advantage to the UK economy post Brexit'.

Agricultural Biotechnology Council briefing paper, 2018.

Cavanagh, Amanda. 'Reclaiming lost calories: tweaking photosynthesis boosts crop yield'. *The Conversation*, January 2019.

Dent, David. *Fixed on Nitrogen: A Scientist's Short Story*. ADG Publishing, 2019.

Doudna, Jennifer. *A Crack in Creation*. Houghton Mifflin, 2017.

Ghislain, Marc et al. 'Stacking three late blight resistance genes from wild species directly into African highland potato varieties confers complete field resistance to local blight races'. *Plant Biotechnology Journal*, 17, 2018: 1119–29.

Hager, Thomas. *The Alchemy of Air: A Jewish Genius, a Doomed Tycoon, and the Scientific Discovery That Fed the World But Fueled the Rise of Hitler*. Crown, 2008.

Lander, Eric. 'The heroes of CRISPR'. *Cell* 164(1), January 2016: 18–28.

Lumpkin, Thomas. 'How a Gene from Japan Revolutionized the World of Wheat: CIMMYT's Quest for Combining Genes to Mitigate Threats to Global Food Security', in Y. Ogihara, S. Takumi and H. Handa (eds). *Advances in Wheat Genetics: From Genome to Field*. Springer, 2015.

Martin-Laffon, Jacqueline, Marcel Kuntz and Agnes Ricroch. 'Worldwide CRISPR patent landscape shows strong geographical biases'. *Nature Biotechnology* 37, 2019: 613–20.

Pappas, Stephanie. 'Irish potato blight originated in South America'. *Live Science*, 3 January 2017.

Reader, John. *Potato: A History of the Propitious Esculent*. Yale University Press, 2009.

Romeis, Jorg et al. 'Genetically engineered crops help support conservation biological control'. *Biological Control* 130, 2019: 136–54.

Van Montagu, Marc. 'It is a long way to GM agriculture'. *Annual Reviews of Plant Biology* 62, 2011: 1–23.

Vietmeyer, Noel. *Our Daily Bread: The Essential Norman Borlaug*. Bracing Books, 2011.

Vogel, Orville. 'Dwarf wheats'. Speech to the Pacific Northwest Historical Society, 1977.

Woodham-Smith, Cecil. *The Great Hunger: Ireland 1845–1849*. Harper and Row, 1962.

5. Low-technology innovation

Devlin, Keith. *The Man of Numbers: Fibonacci's Arithmetic Revolution* (Kindle Locations 840–841). Bloomsbury (Kindle edn).

Guedes, Pedro. 'Iron in Building: 1750–1855 – Innovation and Cultural Resistance'. PhD thesis, University of Queensland, 2010.

Gurley, Bill. 'Money out of nowhere: how internet marketplaces unlock economic wealth'. abovethecrowd.com, 27 February 2019.

Kaplan, Robert. *The Nothing That Is: A Natural History of Zero*. Oxford University Press, 1999.

Levinson, Marc. *The Box*. Princeton University Press, 2006.

McNichol, Ian. *Joseph Bramah: A Century of Invention 1749–1851*. David and Charles, 1968.

Nebb, Adam. 'Why did it take until the 1970s for wheeled luggage to appear when patent applications were being filed in the 1940s?' *South China Morning Post*, 29 June 2017.

Ospina, Daniel. 'How the best restaurants in the world balance innovation and consistency'. *Harvard Business Review*, January 2018.

Petruzzelli, Antonio and Tommaso Savino. 'Search, recombination, and innovation: lessons from haute cuisine'. *Long Range Planning* 47, 2014: 224–38.

Sharky, Joe. 'Reinventing the suitcase by adding the wheel'. *The New York Times*, 4 October 2010.

Spool, Jared. 'The $300 million button'. uie.com, January 2009.

6. Communication and computing

Bail, Christopher. 'Exposure to opposing views on social media can increase political polarization'. *PNAS* 115, 2018: 9216–21.

Handy, Jim. 'How many transistors have ever shipped?' *Forbes*, 26 May 2014.

Isaacson, Walter *The Innovators: How a Group of Hackers, Geniuses, and Geeks Created the Digital Revolution.* Simon and Schuster, 2014.

Johnson, Steven. *Where Good Ideas Come From: The Natural History of Innovation.* Riverhead Books, 2010.

Morris, Betsy. 'A Silicon Valley apostate launches "An inconvenient truth" for tech. *Wall Street Journal*, 23 April 2019.

Newbolt, Henry. *My World As in My Time.* Faber and Faber, 1932.

Pariser, Eli. *The Filter Bubble.* Penguin Books, 2011.

Pettegree, Andrew. *Brand Luther.* Penguin Books, 2015.

Raboy, Marc. *Marconi: The Man Who Networked the World.* Oxford University Press, 2016.

Ribeiro, Marco, Sameer Singh and Carlos Guestrin. '"Why should I trust you?" Explaining the Predictions of Any Classifier'. Proceedings of the 22nd ACM SIGKDD International Conference on Knowledge Discovery and Data Mining 2016, pp. 1135–44.

Silverman, Kenneth. *Lightning Man: The Accursed Life of Samuel F. B. Morse.* Knopf Doubleday, 2003.

Slawski, Bill. 'Just what was the First Search Engine?' seobythesea.com, 2 May 2006.

Thackray, Arnold. *Moore's Law: The Life of Gordon Moore, Silicon Valley's Quiet Revolutionary.* Basic Books, 2015.

The Economist. After Moore's Law. *Technology Quarterly*, 3 December 2016.

7. Prehistoric innovation

Bettinger, R., P. Richerson and R. Boyd. 'Constraints on the development of agriculture'. *Current Anthropology 50*, 2009: 627–31.

Botigué, Laura R., Shiya Song, Amelie Scheu, et al. 'Ancient European dog genomes reveal continuity since the early Neolithic'. *Nature Communications* 8(16082), 2017.

Brown, K. S. et al. 'An early and enduring advanced technology originating 71,000 years ago in South Africa'. *Nature* 491, 2012: 590–93.

Finkel, Meir and Ran Barkai. 'The Acheulean handaxe technological persistence: a case of preferred cultural conservatism?' *Proceedings of the Prehistoric Society* 84, 2018: 1–19.

Gerhart, L. M. and J. K. Ward 'Plant responses to low [CO_2] of the past'. *New Phytologist* 188, 2010: 674–95.

Gowlett, J. A. J. 2016. 'The discovery of fire by humans: a long and convoluted process'. *Philosophical Transactions of the Royal Society* B 371:1696, 2016.

Henrich, J. 'Demography and cultural evolution: how adaptive cultural processes can produce maladaptive losses: the Tasmanian case. *American Antiquity* 69, 2004: 197–214.

Lane, Nick. *The Vital Question: Why is Life the Way It Is?* Profile Books, 2015.

Lovelock, James. *Novacene: The Coming Age of Hyperintelligence*. Penguin Books, 2019.

Mahowald, N., K. E. Kohfeld, M. Hanson, et al. 'Dust sources and deposition during the last glacial maximum and current climate: a comparison of model results with paleodata from ice cores and marine sediments'. *Journal of Geophysical Research* 104, 1999:15,895–15,916.

Marean, C. 'The transition to foraging for dense and predictable resources and its impact on the evolution of modern humans'. *Philosophical Transactions of the Royal Society* B 371(1698), 2016.

McBrearty, S. and A. S. Brooks. 'The revolution that wasn't: a new interpretation of the origin of modern human behavior. *Journal of Human Evolution* 39, 2000: 453–563.

Rehfeld, K., T. Munch, S. L. Ho and T. Laepple, 'Global patterns of declining temperature variability from Last Glacial Maximum to Holocene'. *Nature* 554, 2018: 356–9.

Tishkoff, Sarah A., Floyd A. Reed, Alessia Ranciaro, et al.

'Convergent adaptation of human lactase persistence in Africa and Europe'. *Nature Genetics* 39, 2007: 31–40.

Wrangham, Richard. *Catching Fire*. Profile Books, 2009.

Wrangham, Richard. *The Goodness Paradox*. Profile Books, 2019.

8. Innovation's essentials

Arthur, Brian. *The Nature of Technology: What It is and How It Evolves*. Allen Lane, 2009.

Benjamin, Park. *The Age of Electricity: From Amber-Soul to Telephone*. Scribner, 1886.

Brooks, Rodney. 'The seven deadly sins of AI predictions'. *MIT Technology Review*. 6 October 2017.

Brynjolfson, Erik and Andrew McAfee. *The Second Machine Age: Work, Progress and Prosperity in a Time of Brilliant Technology*. Norton, 2014.

DNA Legal. 'Profile: Sir Alec Jeffreys – the pioneer of DNA testing'. dnalegal.com, 15 January 2015.

Dodgson, Mark and David Gann. *Innovation: A Very Short Introduction*. Oxford University Press, 2010.

Grier, Peter. 'Really portable telephones: costly but coming?' *Christian Science Monitor*, 15 April 1981.

Hammock, Rex. 'So what exactly did Paul Saffo say and when did he say it?'. Rexblog, 15 June 2007.

Harford, Tim. *Fifty Things That Made the Modern Economy*. Little, Brown, 2017.

Harford, Tim. 'What we get wrong about technology'. *Financial Times* magazine, 8 July 2017.

Juma, Calestous. *Innovation and Its Enemies: Why People Resist New Technologies*. Oxford University Press, 2016.

Kealey, Terence and Martin Ricketts. Modelling the industrial revolution using a contribution good model of technical change (unpublished).

Krugman, Paul. 'Why most economists' predictions are wrong'. *Red Herring* Online. 10 June 1998.

McAfee, Andrew. *More from Less: The Surprising Story of How We Learned to Prosper Using Fewer Resources and What Happens Next*. Simon and Schuster, 2019.

Ridley, Matt. *The Evolution of Everything*. HarperCollins, 2015.

Ridley, Matt. 'Amara's Law'. *The Times*, available at Mattridley.co.uk, 12 November 2017.

Schumpeter, Joseph. *Capitalism, Socialism and Democracy*. Harper and Row, 1950.

Wagner, Andreas. *Life Finds a Way*. OneWorld, 2019.

Wall Street Journal. 'Germany's dirty green cars'. 23 April 2019 (editorial).

Wasserman, Edward. 'Dick Fosbury's famous flop was actually a great success'. *Psychology Today*, 19 October 2018.

Wasserman, Edward. A., and Patrick Cullen. 'Evolution of the violin: the law of effect in action'. *Journal of Experimental Psychology: Animal Learning and Cognition*, 42, 2016: 116–22.

West, Geoffrey. *Scale: The Universal Laws of Life and Death in Organisms, Cities and Companies*. Weidenfeld and Nicolson, 2017.

Williams, Gareth. *Unravelling the Double Helix: The Lost Heroes of DNA*. Weidenfeld and Nicolson, 2019.

9. The economics of innovation

Autor, David. 'Why are there still so many jobs? The history and future of workplace automation'. *Journal of Economic Perspectives* 29, 2015: 3–30.

Bush, Vannevar. 'Science: the endless frontier'. US Government, 1945.

Edgerton, David. 'The "Linear Model" Did Not Exist. Reflections on the History and Historiography of Science', in *The Science–Industry Nexus: History, Policy, Implications* (eds. Karl Grandin and Nina Wormbs). Watson, 2004.

Hopper, Lydia and Andrew Torrance. 'User innovation: a novel framework for studying innovation within a non-human context. *Animal Cognition* 22(6), 2019: 1185–90.

Huston, Larry and Nabil Sakkab. 'Connect and develop: inside Procter & Gamble's new model for innovation'. *Harvard Business Review*, March 2006.

Isaacson, Walter. *The Innovators*. Simon and Schuster, 2014.

Jewkes, John. *The Sources of Invention*. Macmillan, 1958.

Kealey, Terence. 'The case against public science'. *Cato Unbound*, 5 August 2013.

Keynes, John Maynard. *Economic Possibilities for Our Grandchildren* (1930), in *Essays in Persuasion*. Norton, 1963.

Loris, Nicholas. 'Banning the incandescent light bulb'. Heritage Foundation, 23 August 2010.

Mazzucato, Mariana. *The Entrepreneurial State*. Anthem Press, 2013.

Mingardi, Alberto. 'A critique of Mazzucato's entrepreneurial state'. *Cato Journal* 35, 2015: 603–25.

Ozkan, Nesli. 'An example of open innovation: P&G'. *Procedia – Social and Behavioral Sciences* 195, 2015: 1496–1502.

Runciman, W. Garry. *Very Different, But Much the Same: The Evolution of English Society since 1714*. Oxford University Press, 2015.

Shackleton, J. R. 'Robocalypse now?' Institute of Economic Affairs, 2018.

Shute, Neville. *Slide Rule: An Autobiography*. House of Stratus, 1954.

Steinbeck, John. 'Interview with Robert van Gelder'. *Cosmopolitan* 18, 1947: 123–5.

Sutherland, Rory. 'Why governments should spend big on tech'. *The Spectator*, 6 July 2019.

Thackeray, William Makepeace. *Ballads and Verses and Miscellaneous Contributions to 'Punch'*. Macmillan, 1904.

von Hippel, Eric. *Free Innovation*. MIT Press, 2017.

Warsh, David. *Knowledge and the Wealth of Nations: A Story of Economic Discovery*. W. W. Norton & Company, 2006.

Worstall, Tim. 'Is there really a "problem" with robots taking our jobs?' CapX.com, 1 July 2019.

10. Fakes, frauds, fads and failures

Alvino, Nicole. 'Theranos: when a culture of growth becomes a culture of scam'. Entrepreneur.com, 22 May 2019.

Carreyrou, John. *Bad Blood: Secrets and Lies in a Silicon Valley Startup*. Pan Macmillan, 2018.

Dennis, Gareth. 'Don't believe the hype about hyperloop'. *Railway Gazette*, 14 March 2018.

Rowan, David. *Non-Bullshit Innovation*. Bantam Press, 2019.

Stern, Jeffrey. 'The $80 million fake bomb-detector scam – and the people behind it'. *Vanity Fair*, 24 June 2015.

Stone, Brad. *The Everything Store: Jeff Bezos and the Age of Amazon*. Little, Brown, 2013.

Troianovski, Anton and Sven Grundberg. 'Nokia's bad call on smartphones'. *Wall Street Journal*, 11 July 2012.

11. Resistance to innovation

Behrens, Dave. 'The Tabarrok Curve: a call for patent reform in the US'. *The Economics Review at NYU*, 13 March 2018.

Cerf, Vint, Tim Berners-Lee, Anriette Esterhuysen, et al. 'Article 13 of the EU Copyright Directive threatens the internet'. Letter to Antonio Tajani MEP, President of the European Parliament, 12 June 2018.

Chisholm, John. 'Drones, dangerous animals and peeping Toms: impact of imposed vs. organic regulation on entrepreneurship, innovation and economic growth'. *International Journal of Entrepreneurship and Small Business* 35, 2018: 428–51.

Dumitriu, Sam. 'Regulation risks making Big Tech bigger'. CapX.com, 27 November 2018.

Erixon, Fredrik and Björn Weigel. *The Innovation Illusion: How So Little is Created by So Many Working So Hard*. Yale University Press, 2016.

Hazlett, Tom. *The Political Spectrum: The Tumultuous Liberation of Wireless Technology, from Herbert Hoover to the Smartphone*. Yale University Press, 2017.

Hazlett, Tom. 'We could have had cellphones four decades earlier'. *Reason Magazine*, July 2017.

Heller, Michael. *The Gridlock Economy: How Too Much Ownership Wrecks Markets, Stops Innovation, and Costs Lives*. Basic Books, 2008.

Juma, Calestous. *Innovation and Its Enemies: Why People Resist New Technologies*. Oxford University Press, 2016.

Lindsey, Brink and Steve Teles. *The Captured Economy*. Oxford University Press, 2017.

Lynas, Mark. *Seeds of Science: Why We Got It So Wrong on GMOs*. Bloomsbury, 2018.

Mcleod, Christine. *Inventing the Industrial Revolution: The English Patent System 1660–1800*. Cambridge University Press, 1988.

Paarlberg, Robert. 2014. 'A dubious success: the NGO campaign against GMOs'. *GM Crops & Food* 5(3), 2014: 223–8.

Porterfield, Andrew. 'Far more toxic than glyphosate: copper sulfate, used by organic and conventional farmers, cruises to European reauthorization'. Genetic Literacy Project, 20 March 2018.

Tabarrok, Alex. *Launching the Innovation Renaissance*. TED Books, 2011.

Thierer, Adam. *Permissionless Innovation: The Continuing Case for Comprehensive Technological Freedom*. Mercatus Center, George Mason University, 2016.

Thierer, Adam. *Permissionless Innovation and Public Policy: A 10-Point Blueprint*. Mercatus Center, George Mason University, 2016.

Zaruk, David. 'Christopher Portier – well-paid activist scientist at center of the ban-glyphosate movement'. Genetic Literacy Project, 17 October 2017.

12. An innovation famine

Erixon, Fredrik and Björn Weigel. *The Innovation Illusion: How So Little is Created by So Many Working So Hard*. Yale University Press, 2016.

Lindsey, Brink and Steve Teles. *The Captured Economy.*
  Oxford University Press, 2017.

Topol, Eric. *Deep Medicine: How Artificial Intelligence
  Can Make Healthcare Human Again.* Basic Books, 2019.

Torchinsky, Jason. 'Here's how fast cars would be if they
  advanced at the pace of computers'. Jalopnik.com,
  2 March 2017.

Withdrawn from Stock

EPIGRAPH SOURCES

Introduction: Schumpeter, Joseph. *Capitalism, Socialism and
  Democracy.* Harper and Row, 1950.

1: Attributed to Peter Drucker.

2: Pylarini, Jacob. 'A new and safe method of communicating the
  small-pox by inoculation, lately invented and brought into use'.
  Philosophical Transactions of the Royal Society 29:347, 1716.

3: Attributed to Henry Ford.

4: Reader, John. *Potato: A History of the Propitious Esculent.*
  Yale University Press, 2009.

5: Kaplan, Robert. *The Nothing That Is: A Natural History of
  Zero.* Oxford University Press, 1999.

6: *The Economist.* After Moore's Law. *Technology Quarterly,*
  12 March 2016.

7: Haldane, J.B.S. 'Daedalus, or science of the future'. Lecture
  to the Heretics Society, Cambridge, 4 February 1923.

8: Jefferson, Thomas. Letter to Joseph Willard, 24 March 1789.

9: Steinbeck, John. Interview with Robert van Gelder (April
  1947), as quoted in *John Steinbeck: A Biography* by Jay
  Parini. William Heineman, 1994.

10: Bezos, Jeff. Speaking at the Amazon re:MARS conference,
  Las Vegas, June 2019.

11: Petty, William. *Treatise on Taxes and Contributions*
  (pp. 113–14), 1662.

12: Thiel, Peter. Speaking at the Yale School of Management,
  27 April 2013.

ACKNOWLEDGEMENTS

Many people have helped me to write this book and I am deeply grateful to them all. Since one of my themes is that innovation is a more collaborative exercise than usually recognized, I am glad to admit that I am the sum, or product, of the many influences on me.

Those who helped my thinking in writing and in conversation over the past few years include Reuben Abraham, Jun Arima, Alastair Balls, Isabel Behncke, John Bell, Robert Boyd, John Burn, Gabriel Calzada, Douglas Carswell, Bill Casebeer, Monika Cheney, John Chisholm, John Constable, Frederic Darriet, David Dent, Susan Dudley, Natascha Engel, Fredrik Erixon, Fiona Fell, Ian Fells, Greg Finch, George Freeman, Dominic Frisby, Gordon Getty, Josh Gilder, Oliver Goodenough, Ian Gregory, Bill Gurley, Dan Hannan, Tom Hazlett, David Hill, Lydia Hopper, Anton Howes, the late Calestous Juma, Terence Kealey, Michael Kelly, Mark Littlewood, Kelly-Jo Macarthur, Brian Mannix, Mike Mayerhof, Andrew Mayne, Kevin McCabe, Deirdre McCloskey, Alberto Mingardi, Julian Morris, Jon Moynihan, Jesse Norman, Gerry Ohrstrom, Kendra Okonski, Owen Paterson, Ryan Phelan, Pete Richerson, Paul Romer, David Rose, Max Roser, Bartlett Russell, Alan Rutherford, Vaclav Smil, Nick Steinsberger, Taishi Sugayama, Rory

Sutherland, Eors Szathmary, Andrew Torrance, Liz Truss, Marian Tupy, Magatte Wade, Edward Wasserman, Richard Webb, Bruce Whitelaw, Candida Whitmill, Matthew Willetts, Richard Wrangham, Chris Wright, David Zaruk and many more.

Over the years, I have interviewed or spoken to many great innovators and entrepreneurs, from whom I have drawn ideas about this topic. They include Jeff Bezos, Stewart Brand, James Dyson, Alec Jeffreys, James Lovelock, Eric Schmidt, Peter Thiel, Tony Trapp, James Watson and Mark Zuckerberg.

I test-drove some of the ideas in this book in columns for the *Wall Street Journal* and *The Times* and am grateful to my editors, Gary Rosen and Mike Smith, for their help and support.

Colleagues in the House of Lords have often proved rich seams of new thoughts. I won't try to name them all but I am especially indebted to Naren Patel, chairman of the Science and Technology Select Committee, and Tim Clement-Jones, chairman of the Artificial Intelligence Select Committee, for the chance to serve under their expert chairmanship and use the evidence sessions of inquiries to deepen my understanding of innovation.

I am especially grateful to John Constable, whose ideas very much infuse this book, especially his thinking on the notion of improbability, and who has been a sounding board and editor both. Also to Derick Bellamy for help with social media, and to Sarah Thickett for help with editing.

My agents, Felicity Bryan and Peter Ginsberg, and my editors, Louise Haines and Terry Karten, have been enthusiastic champions of this book and wise readers of early drafts for which I thank them sincerely.

Last but most, great thanks to my wife, Anya, a profound thinker about discovery and invention, and to my children for helpful conversations and delightful company.

INDEX